MW01345681

Philosophy of Olfactory Perception

Andreas Keller

Philosophy of Olfactory Perception

Andreas Keller
Astoria, New York, USA

ISBN 978-3-319-33644-2 ISBN 978-3-319-33645-9 (eBook)
DOI 10.1007/978-3-319-33645-9

Library of Congress Control Number: 2016960679

Printed on acid-free paper

This Palgrave Macmillan imprint is published by Springer Nature
The registered company is Springer International Publishing AG
The registered company address is: Gewerbestrasse 11, 6330 Cham, Switzerland

Introduction: Why Study Philosophy of Olfactory Perception?

Plato wrote that smell is of a "half-formed nature" and that not much can be said about it. Two thousand years later, Immanuel Kant identified smell as the "most ungrateful" and "most dispensable" of the senses. Many contemporary philosophers seem to agree and olfaction is therefore dismissed or ignored in most philosophical accounts of perception. The goal of this book is to show how this omission distorts our understanding of what perception is.

I am not the first to realize the potential of opening up perceptual philosophy to the non-visual modalities. Bill Lycan wondered "how the philosophy of perception would be different if smell had been taken as a paradigm rather than vision" (Lycan 2000, p. 273). The answer, as I will show here, is that the philosophy of perception would be *very* different if it were based on olfaction. Considering olfaction reveals that many of the most basic concepts of the philosophy of perception are based on peculiarities of visual perception that are not found in other modalities.

The focus on olfaction (Fig. 1), which is a simple and well-understood sensory modality, is the main difference between this book and other works on perception. In addition to the emphasis on smells, the account of perception presented here also differs from most other approaches in its appreciation of the fact that perception is an evolved capacity. As such, perception can only be understood within an evolutionary framework and developing a theory of perception therefore requires collaboration

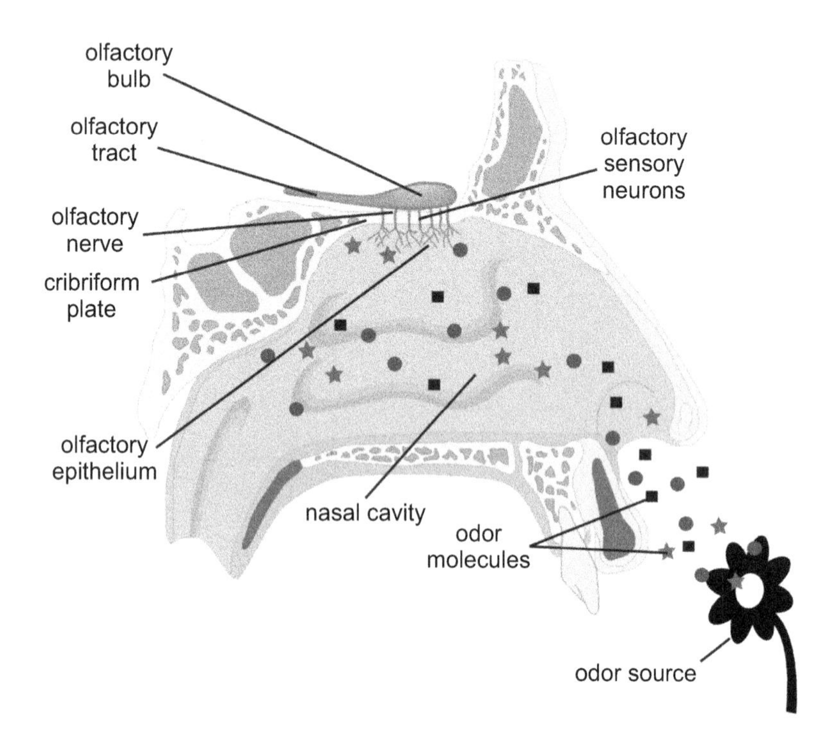

Fig. 1 The olfactory system. Odor sources give off a mix of different types of odor molecules (represented in this figure by *rectangles*, *stars*, and *filled circles*). During sniffing, the odor molecules enter the nasal cavity. At the top of the nasal cavity is the olfactory epithelium, which contains the odor-sensitive endings of the olfactory sensory neurons. The olfactory sensory neurons form the olfactory nerve, which connects the nasal cavity with the brain through small holes in the skull (in the cribriform plate). In the brain, the olfactory sensory neurons terminate in the olfactory bulb, the first processing center for odor information. The olfactory bulb is connected to other brain regions through the olfactory tract

between different disciplines. The questions addressed in this book are the questions of perceptual philosophy. The answers to these questions combine insights and results from a variety of disciplines, including philosophy, neuroscience, cognitive science, and psychology.

This book is divided into four parts: "Perceptual Qualities", "Percepts", "Olfaction and Cognitive Processes", and "Consciousness". In the first part, I will discuss topics related to perceptual qualities.[1] Philosophically important issues related to perceptual qualities that will be addressed are the nature of perceptual qualities, modality individuation, and the possibility of third-person access to perceptual qualities. I will discuss perceptual qualities as they are revealed by behavioral or psychophysical experiments in which different behaviors are shown in response to different stimuli. In the first chapter, I will suggest a strategy to arrange the perceptual qualities perceived by an individual perceiver in a perceptual space that reflects the similarity relations between the perceptual qualities. In the second chapter, the problem of comparing perceptual qualities between different perceivers is addressed. I will suggest that it is possible to compare perceptual qualities perceived by two different perceivers when the two perceivers' perceptual spaces can be registered. Registration of perceptual spaces is the process of transforming one of the spaces (the target) so that it fits into the same coordinate system as the other space (the reference).

Typically, perception is much more complex than a series of discriminable perceptual qualities. In the second part, these complexities will be addressed. In Chap. 3, I will argue that olfactory perception has no spatial structure. In modalities in which perception has a rich spatial structure, perception is often thought of as the perception of objects that are spatially extended and bounded. I will apply common notions of perceptual objects to olfaction and show that odor perception is not the perception of objects. Chapter 4, which is a bit of a digression, will address the question of the evolutionary function of perception. I will argue that it is the function of perception to guide behaviors and defend this idea against alternative proposals. The general goal of the second part of this book is to demonstrate that many of the complexities that are taken to be an integral part of perception in other modalities are absent in olfactory perception. These complexities therefore play no role in an account of perception that is based on olfaction.

Perception and non-perceptual cognitive processes are not always clearly separated. Perceptual systems coevolved with cognitive systems that process perceptual information and motor systems that execute an organism's behavior. Perception depends on these systems for being useful to the organism. Perception itself has no adaptive advantages unless it results in a stimulus-dependent modification of behavior.[2] In the third part, I will address the connections between perceptual systems and non-perceptual cognitive systems. A review of the availability of olfactory information for cognitive systems in Chap. 5 will show that the connections between perceptual and cognitive processes differ between modalities. The olfactory system evolved to efficiently provide input to emotional systems, but not to language systems. In Chap. 6, I will discuss input to the olfactory system from other parts of the mind. Specifically, I will investigate the evidence for cognitive penetration and crossmodal perception in olfaction.

Throughout the first three parts of this book, perception is discussed with respect to its ability to guide behaviors and to make information available to cognitive processes. In many instances, the perception is conscious. However, the conclusions drawn in Parts 1 to 3 are supposed to hold for conscious and non-conscious instances of perception. The fourth part will address the differences between conscious and non-conscious olfactory perception. This discussion of consciousness is more speculative than the rest of the book. Based on olfactory perception, I will argue in Chap. 7 for an important role of attention in conscious processes and in Chap. 8 that the function of conscious brain processes is to guide behaviors in complex situations.

Notes

1. A variety of terms are used for what I will call "perceptual qualities". Daniel Dennett lists "raw feels", "sensa", "phenomenal qualities", "intrinsic properties of conscious experiences", "qualitative content of mental states", and "qualia". Dennett, D. C. (1991). *Consciousness Explained.* Boston, Little, Brown and Company (p. 372). Each term tacitly imports assumptions about the phenomena that are labeled. I mean by "perceptual qualities" the mental qualities that are different in cases in which two colors, smells, tastes, and so on can be behaviorally discriminated.
2. As an illustration of the uselessness of perception for perception's sake, consider sea squirts. Many species of sea squirts have a free-swimming larval form and a sedentary adult form. During the development from the free-swimming to the adult form, both the muscular system and the nervous system degenerate. The function of the brain is to guide behaviors, and in the absence of behavioral

options there is no need for a brain. Sensing a predator is only adaptive for an organism that has behavioral strategies for predator avoidance.

References

Dennett, D. C. (1991). *Consciousness explained.* Boston: Little, Brown and Company.

Lycan, W. G. (2000). The slighting of smell. In N. Bhushan & S. Rosenfeld (Eds.), *Of minds and molecules: New philosophical perspectives on chemistry.* (pp. 273–289). Oxford: Oxford University Press.

Acknowledgments

Too many people to mention have helped me in countless ways during the writing of this book. Most important for the success of this project were the love, support, and encouragement from my wife Dara Mao. My teachers, friends, and colleagues Jesse Prinz, Peter Godfrey-Smith, David Rosenthal, Jessica Keiser, Svetlana Novikova, Ann-Sophie Barwich, and Benjamin Young have read all or parts of earlier versions of this manuscript and helped me improving it. I am very thankful and hope to get a chance to return the favor some time. Of course, all the views presented here and all the mistakes throughout the book are mine.

Contents

Part 1

Perceptual Qualities

The basic building blocks of perception are perceptual qualities with no spatial or temporal structure, like the redness of a tomato or the characteristic smell of a rose.[1] These perceptual qualities are mental qualities that, in humans, are usually considered to be conscious. The redness and smell of roses are often consciously experienced by the perceiver. Although this is the most familiar way in which we encounter them, perceptual qualities are not necessarily consciously experienced (Rosenthal 2010, 2016 Young et al. 2014). The fact that perceptual qualities can be either conscious or non-conscious is best illustrated by cases in which behavioral decisions are made based on perceptual qualities in the absence of consciousness. For example, at very low odor concentrations people often report verbally that they cannot detect an odor, although they discriminate stimuli successfully when they are asked to make a choice (Sobel et al. 1999). At slightly higher concentrations, when the sensory information is processed consciously, two odors are discriminated based on differences in their perceptual qualities. It is parsimonious to assume that the same two odors at lower concentrations are also discriminated based on differences in their perceptual qualities, even though the information is not consciously processed. Denying the existence of non-conscious perceptual qualities would require two theories of perception. One theory would have to explain how we distinguish a green object from a red object based on the different consciously perceived perceptual qualities associated with the objects. The second theory would have to explain how we discriminate stimuli without using differences between perceptual qualities in cases of stimulus-dependent behaviors in the absence of conscious processes. If the existence of non-conscious perceptual qualities is admitted, a single theory of perceptual discrimination based on perceptual qualities can explain all cases of perceptual discrimination.

To avoid the need for two different theories of perception depending on whether the perception is conscious or not, one could also deny the existence of non-conscious perception. If all perception is conscious, then there can be no non-conscious perceptual qualities. However, empirical evidence suggests that humans show stimulus-dependent behaviors in the absence of conscious processing. In olfaction, evidence for non-conscious perception comes from experiments that have shown that odors can have specific behavioral effects regardless of whether the olfactory information is processed consciously or not. For example, sniffs are shorter and shallower when an unpleasant odor is encountered than when a pleasant odor is encountered. This effect is independent of conscious experience (for a detailed treatment of this topic in olfaction, see Young 2014). Influences of unconscious perceptual qualities on behaviors are also found in other modalities. In vision, masked priming effects, for example, depend on the features of the masked (e.g., not consciously perceived) stimulus (for reviews, see Cruse et al. 2007; Hallett 2007). Evidence for non-conscious perception has accumulated over the last decade. However, the complete absence of consciousness is difficult to prove and some researchers are not convinced that non-conscious perception exists in humans. The account of perceptual qualities presented here does not depend on the existence of non-conscious perception or non-conscious perceptual qualities. Instead, whether perception is conscious or not is irrelevant because the account is based on the outcome of behavioral or psychophysical experiments alone. The results of this analysis apply to perception in humans, birds, bacteria, robots, or any other system that shows differential responses to different physical stimuli.

Perceptual qualities are the building blocks of perception and all instances of perception involve perceptual qualities. All visual perception, for example, involves colors, although sometimes they are achromatic colors (black, white, shades of gray). Similarly, all olfactory perception involves smells. There are myriads of colors and tones and smells.

Attempts to bring order to the diversity of perceptual qualities go back to the earliest days of philosophy. The most promising approach has been to arrange perceptual qualities relationally according to their similarities in a multidimensional mathematical space (Goodman 1951; Clark 1993; Matthen 2005; Rosenthal 2014; Young et al. 2014). Using this approach, researchers have arranged all colors in a three-dimensional perceptual space (Fig. 1.1), with the dimensions hue, brightness, and saturation. All tones have been arranged in a two-dimensional space according to their pitch and loudness.[2]

Arranging things depending on their relations to each other has been a successful approach to order biological diversity outside of perceptual research. The most prominent example is the arrangement of all living things based on their evolutionary relationships in a phylogenetic tree. One of the reasons the phylogenetic tree is so important for evolutionary biology is that it includes *all* living things and the relations between them. The perceptual space that I suggest similarly includes *all* perceptual qualities, regardless of their modality. This is the philosophically most important difference between the approach presented here and previous approaches which my suggestions are heavily based on (Goodman 1951; Clark 1993; Matthen 2005; Rosenthal 2014; Young et al. 2014). Previous approaches usually focused on constructing perceptual spaces for individual modalities (color space, smell space, etc.). It is my hope that a perceptual quality space that includes all perceptual qualities is an interesting alternative. In the first chapter, I will outline an approach to construct such a perceptual quality space that individuates perceptual qualities by arranging them according to their similarities. Each perceptual quality is individuated by its unique position in the perceptual quality space. I will suggest using triadic comparisons to determine relative similarity between three stimuli to arrange the stimuli in a multidimensional space. I will explain and justify this method and then speculate about the features that the resulting perceptual quality space can be expected to have.

The obvious disanalogy between the phylogenetic tree and the perceptual quality space is that there is only one phylogenetic tree. On the other hand, there are many different perceptual quality spaces. Individuals with red-green color blindness have a different perceptual space from normal-sighted individuals. Furthermore, an individual's perceptual quality space can change depending on experiences and changes in the sensory systems. In the second chapter, I will first discuss this diversity of perceptual quality spaces to set up the problem of comparing perceptual qualities between different perceivers. I will then show that, under certain circumstances, individuating perceptual qualities without reference to subjective experience through quality spaces makes it possible to compare perceptual qualities that are perceived by different perceivers.

Notes

1. As will be discussed in Part 2, these perceptual qualities often occur in complicated spatial and temporal arrangements. However, for the treatment in the first part, I will abstract away from spatial and temporal structures.
2. I am ignoring timbre in the context of perceptual spaces. Timbre is the quality of sound that allows us to discriminate between the sound made by a piano and by a guitar when they have the same pitch and loudness. Timbre has been called, "the psychoacoustician's multidimensional waste-basket category for everything that cannot be labeled pitch or loudness." McAdams, S. and A. Bregman (1979). "Hearing musical streams." *Computer Music Journal* **3**(4): 26–43 (p. 34). The rationale for ignoring timbre is that it is determined by temporal characters of the sound, like its onset and time envelope. The perceptual spaces discussed in the first part of this book are meant to be arrangements of perceptual qualities, abstracted away from their temporal and spatial structure. Abstracting away from temporal structure is more difficult in audition than in other modalities. Pitch, which is usually included as a dimension of tone spaces, is determined by the frequency of a sound and therefore also dependent on the sound's temporal features. To include pitch but exclude temporal features that contribute to timbre is admittedly arbitrary. I do not think that expanding the tone space into a sound space by including the features that contribute to timbre would have important theoretical consequences beyond increasing complexity. However, this question may be worth more consideration.

References

Clark, A. (1993). *Sensory qualities*. Cambridge: Cambridge University Press.

Cruse, H., Kälberer, H., et al. (2007). Sensorimotor supremacy: Investigating conscious and unconscious vision by masked priming. *Advances in Cognitive Psychology, 3*(1), 257–274.

Goodman, N. (1951). *The structure of appearance*. Cambridge: Harvard University Press.

Hallett, M. (2007). Volitional control of movement: The physiology of free will. *Clinical Neurophysiology, 118*(6), 1179–1192.

Matthen, M. (2005). *Seeing, doing, and knowing*. Oxford: Oxford University Press.

McAdams, S., & Bregman, A. (1979). Hearing musical streams. *Computer Music Journal*, 3(4), 26–43.

Rosenthal, D. M. (2010). How to think about mental qualities. *Philosophical Issues: Philosophy of Mind, 20*, 368–393.

Rosenthal, D. (2014). Quality spaces and sensory modalities. In P. Coates & S. Coleman (Eds.), *The nature of phenomenal qualities: Sense, perception, and consciousness*. Oxford: Oxford University Press.

Rosenthal, D. M. (2016). Quality spaces, relocation, and grain. In J. R. O'Shea (Ed.), *Wilfrid Sellars and his legacy*. Oxford: Oxford University Press.

Sobel, N., Prabhakaran, V., et al. (1999). Blind smell: Brain activation induced by an undetected air-borne chemical. *Brain, 122*(Pt 2), 209–217.

Young, B. D. (2014). Smelling phenomenal. *Frontiers in Psychology, 5*, 713.

Young, B. D., Keller, A., et al. (2014). Quality space theory in olfaction. *Frontiers in Psychology, 5*, 1.

1

Perceptual Quality Space

In this chapter, I will focus on how to construct the perceptual quality space for a given individual at a given point in time. I will not concern myself with questions about how stable this space is over time or how similar one individual's space is to another's. These topics will be addressed in Chap. 2. In this chapter, I will only attempt to develop and defend a strategy that arranges perceptual qualities in a way that reliably reflects the similarities between how they are perceived by an individual perceiver. Traditionally, the problem of putting order to the bewildering diversity of perceptual qualities has been addressed in two steps. In a first step, each perceptual quality was assigned to a sensory modality. In the second step, the perceptual qualities of a given modality were arranged in a multidimensional space according to the similarities between the perceptual qualities. The result of this two-step process is a list of modalities and separate perceptual quality spaces for each modality. What is missing from this result is an explanation of similarities between stimuli in different modalities. It can be argued that there are no stable, meaningful similarity relations between tones, colors, tastes, and smells. However, some observations suggest that perceived similarities between perceptual qualities do not respect traditional modality boundaries. For example,

A. Keller, *Philosophy of Olfactory Perception*,
DOI 10.1007/978-3-319-33645-9_1

sugar, which is an odorless tastant, is consistently experienced to be more similar to the tasteless odorant vanillin than to salt, another odorless tastant (Rankin and Marks 2000). If all we have is a list of modalities and the structure of the perceptual quality spaces for each of the modalities, then we cannot account for this fact. To address this problem, I propose a one-step strategy of arranging all perceptual qualities regardless of their modality in a multidimensional space based on their similarity relations. The resulting space will be relational "all the way down".

1.1 Constructing an Exhaustive Perceptual Quality Space

An exhaustive perceptual quality space is a space in which all smells, colors, tones, tastes, and other perceptual qualities are arranged. Perceptual quality spaces consist of multidimensional coordinate systems in which different locations represent different perceptual qualities. Partial perceptual quality spaces for tones and colors (Fig. 1.1) have already been established. As far as these partial spaces reflect similarity relations

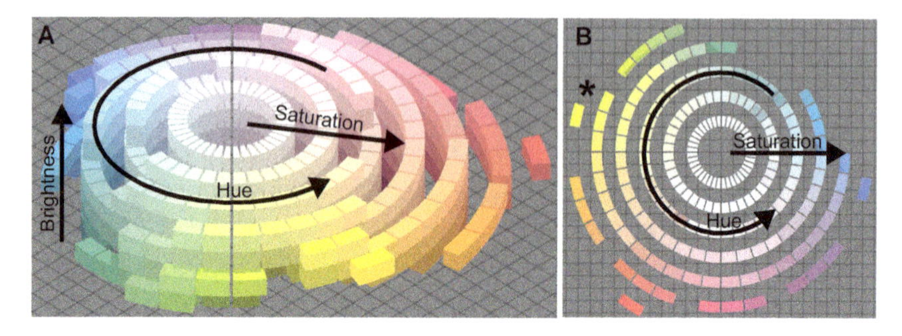

Fig.1.1 The perceptual color quality space. (**a**) The perceptual color quality space is three-dimensional and asymmetric. By convention, the hue dimension is represented as a *closed circle*. The vertical dimension is brightness and saturation increases from the inside of the circle formed by the hue dimension. (**b**) A section through the perceptual color quality space at a given brightness. The *asterisk* marks the position of a color that does not exist in normal perception. A perceptual quality with this combination of hue, brightness, and saturation is not normally perceived by humans

between the perceptual qualities of a modality, they will become part of the exhaustive perceptual space. A possible way to arrive at an exhaustive perceptual space is to first individuate the modalities, then construct perceptual spaces for each modality, and then arrange these perceptual spaces in a modality space. A modality space is a multidimensional space in which modalities are arranged according to the similarities between them (Keeley 2002; Gray 2013; Macpherson 2015). This strategy does not address the problem I am interested in; the resulting modality space of perceptual spaces will not explain why sugar is more similar to vanillin than to salt (Rankin and Marks 2000). To get an explanation for this observation, a space of modalities is not sufficient. Instead, one needs a space of perceptual spaces, in which the perceptual spaces can have different orientations with respect to each other. In short, one needs a space in which perceptual qualities belonging to different modalities are arranged according to similarity relations between them. To achieve this, I propose a one-step procedure of arranging all perceptual qualities in an exhaustive perceptual space based on psychophysical measures of similarity.

Determining Similarity Between Perceptual Qualities

It is difficult to overestimate the importance of similarity judgments for an organism's survival. Similarity judgments play a much more pronounced role in perception-based behaviors than discrimination. Discrimination is not very useful because we almost never encounter indiscriminable things. An apple can perceptually be discriminated from all (or most) other apples perceptually. It can also be discriminated from all bananas and all other things in the world. The capacity to discriminate a large, ripe, intact apple from another large, ripe, intact apple is not a useful evolutionary adaptation. What is useful is to be able to judge that one large, ripe, intact apple is similar to another large, ripe, intact apple. This allows us to respond similarly to things that are perceived similarly. That we are able to make very fine discriminations is a consequence of a perceptual system that evolved to judge similarities. As Quine wrote, "surely there is nothing more basic to thought and language than our sense of similarity; our sorting of things into kinds" (Quine 1970).

Experimentally, perceived similarities can be revealed in different ways (for a detailed discussion of the advantages and disadvantages of different strategies in olfaction, see Wise et al. 2000). Many of these experiments can be performed with humans as well as with other animals, and often the results from different species are similar (Chen and Gerber 2014). One reliable psychophysical method for determining similarity relations between stimuli is triadic comparison. In triadic comparisons, three stimuli are presented to subjects who are instructed to pick the two stimuli that are the most similar (MacRae et al. 1990, 1992). Other methods for determining perceived similarities between stimuli include direct ratings of the similarity of two odors, and measuring similarity between verbal descriptors that are applied to odors (Wise et al. 2000). Each of these methods has advantages and disadvantages and using different psychophysical methods to determine stimulus similarity will undoubtedly lead to differently structured perceptual spaces. However, this dependence on the procedure used for its construction does not diminish the explanatory power of the perceptual quality space. This can be illustrated by comparison to phylogenetic trees. Phylogenetic trees can be built based on skeletal anatomy, genomic sequences, or a combination of both. The resulting trees will be different. Mostly the differences will be small, but sometimes they will be consequential. The extent evidence is unlikely to be sufficient to resolve all uncertainties about shifts in populations that occurred hundreds of millions of years ago. There is a phylogenetic tree that accurately reflects all relationships between all living things. However, it is unlikely that we will ever achieve great certainty about every detail of the relationships represented by that tree. This uncertainty does not diminish the importance of phylogenetic trees for biology. Similarly, there are, for a given perceiver at a given time, true similarity relations between all perceptual qualities. Different psychophysical methods or combinations of these methods will result in slight differences in how these relations are represented in a perceptual quality space. However, the importance of the perceptual quality space is not undermined by ambiguities due to different methods of constructing it.

Different methods can be used to establish similarity relations between stimuli. Using triadic comparisons has several advantages over other methods. One advantage of this method is that it requires only ordinal judgments. The similarity between stimuli does not have to be quantified, which is a cognitively very demanding task. Another advantage is that, as forced-choice tests, triadic comparisons are independent of conscious perception of similarities. Subjects that report that all three stimuli are the same may still perform non-randomly when asked to pick the two that are the most similar. A further advantage is that triadic comparisons, or equivalent stimulus generalization experiments, can be performed by all species and in all modalities. Presenting three stimuli in a row and asking subjects to group them according to similarity is possible in all modalities. Other methods to gain information about the similarity relations between stimuli are based on presenting two stimuli, for example, colors, simultaneously. These types of experiments are only possible for modalities in which different stimuli can be perceived simultaneously.

A complication for all attempts to construct a perceptual quality space is the existence of perceptual qualities that are not directly associated with physical stimuli. There are, for example, "impossible" perceptual color qualities that can be only produced by combining the perception of the afterimage of one color with the perception of another color. If one looks at a pale blue-green surface and immediately afterwards at a maximally saturated orange surface, for example, the combination of the orange afterimage and the orange perception results in the perception of a hyperbolic orange that is more ostentatiously orange than every orange normally seen (Churchland 2005). Producing such color qualities poses a practical problem for constructing the perceptual space. Instead of simply presenting the stimulus, the whole sequence of stimuli that reliably leads to the perceptual quality has to be presented. However, as long as the perceptual qualities are predictably inducible, the problem is only practical. If some perceptual qualities are *only* perceived in hallucinations, and this perception is independent of a stimulus and impossible to predict, then these qualities cannot be captured by the method I propose.

Relation Between Similarity and Discriminability

Similarity-based psychophysics are not usually used to construct perceptual spaces. More commonly, perceptual quality spaces are constructed based on subjects' ability to discriminate stimuli which can be determined using the just-noticeable-difference method (Goodman 1951; Clark 1993;

Matthen 2005; Rosenthal 2014; Young et al. 2014). The just-noticeable difference between two stimuli can be determined by forced-choice-discrimination tasks. The experiments start with two discriminable stimuli. The subject is asked to discriminate the stimuli and when they do so reliably, the physical properties of one of the two stimuli are gradually made more similar to the physical properties of the other stimulus. When the point is reached at which they can no longer be discriminated, the just-noticeable difference between the two stimuli has been discovered. If the two stimuli cannot be made to be indiscriminable, then they belong to different modalities (Rosenthal 2014). The just-noticeable-difference method results in a color space, a smell space, and several other quality spaces. However, it does not produce information about the similarities between stimuli in different modalities (because each tone can be discriminated from each taste).

The results of forced-choice-discrimination tasks that are used in the just-noticeable-difference method are a subset of the results of triadic comparisons. When the subject in a triadic comparison task is presented with three stimuli, two of which are identical, then the grouping of the stimuli depends on whether the subject can discriminate the two identical stimuli from the third stimulus. When the difference between the stimulus that is presented in duplicate and the third stimulus is just noticeable, the subject will rate the two identical stimuli as being the most similar pair. When the third stimulus is then altered so that it can no longer be discriminated from the two identical stimuli, then the responses will be random, with each pair being identified as the most similar pair with equal likelihood. In this way, triadic comparisons provide information about just-noticeable differences.

Triadic comparisons reveal information about just-noticeable differences. In addition to the information whether two stimuli can be discriminated, triadic comparisons also produce information about the relative perceptual similarity of discriminable stimuli pairs. This information is necessary to construct an exhaustive perceptual quality space that accounts for similarity relations between perceptual qualities in different modalities.

Despite the additional information that can be obtained through similarity-based psychophysics, psychophysicists are usually more comfortable with discrimination tasks than with similarity-based psychophysics. One reason why discrimination tasks are often preferred to similarity judgments is that discrimination tasks are the paradigmatic example of psychological measures of performance. Similarity judgments, on the other hand, are often considered to be based on the report of mental content rather than on performance (Wise et al. 2000). However, when triadic comparisons are used to determine similarity relations, this is not the case. There are objectively correct facts about perceptual similarities and similarity tasks are a test of the ability to identify them. The length of a three-inch nail is more similar to the length of a four-inch nail than to the length of an eight-inch nail. Similarly, of the distances between three perceptual qualities in a perceptual space, one distance is in most cases shorter than the other two distances. To deny that there are objectively correct answers to triadic comparisons is to deny that the perceptual qualities involved in the triadic comparison can be arranged according to similarity in a perceptual quality space.

I said that in triadic comparisons *in most cases* one of the three distances between the three perceptual qualities is shorter than the other two. One situation in which no distance is shorter than the other two is when all three distances are the same. They can all be zero, in which case the three stimuli are indiscriminable, or all three distances can have the same non-zero length. In this case, the stimuli can be discriminated, but the distances between the three stimuli in stimulus space are indiscriminable. This situation is most often encountered when the three stimuli in the triadic comparison are very dissimilar. With increasing dissimilarity, differences in similarity become more difficult to detect. This is presumably because similarity judgments are most useful when they concern relatively similar things. The function of similarity judgments is to cluster, generalize, or categorize things, and when things are so dissimilar that they clearly are not in the same cluster, the capacity to compare similarities is not advantageous. This failure to perform in triadic comparisons of three very different stimuli is not a bug, but a feature of triadic comparisons. Whether people can or cannot compare the similarities between stimuli has to be found out experimentally for individual cases. I predict that a triadic comparison between the color red, cold temperature, and bitter taste will not reveal anything about the similarity relations of these perceptual qualities. If this is the case, then the data from this triadic comparison will not contribute to the construction of the perceptual quality space. However, I may be wrong and performing the experiment may reveal that people group cold temperature with bitter taste. If this is the case, then the data about the relative similarity between the three stimuli can be used in the construction of the perceptual quality space.

Another common objection to using similarity judgments in psychophysics is that similarity can be judged along different criteria (Wise et al. 2000). Imagine, for example, a triadic comparison of dark blue, light blue, and dark green. Some subjects may use brightness as a criterion whereas others may use hue. Imagine, as suggested by Matthen (2005, p. 131), that you are asked whether the USA, Russia, or China is more similar to Canada. Your answer will depend on the context in which the question is asked, and depending on the context, you will privilege different criteria (climate, political system, size, population, etc.) in answering. However, there is an important methodological difference between judging similarities of countries and of perceptual qualities. Unlike countries, perceptual qualities can usually be altered along a single criteria (or dimension) and then similarity relations between perceptual qualities that differ in many different criteria can be revealed. It is difficult to decide whether the USA or Russia is more similar to Canada. However, it is easy to decide whether the actual USA or a fictional version of it that differs only in that it has a colder climate is more similar to Canada. That similarities can be judged along different criteria is therefore not as big a problem for assessing similarities between perceptual qualities than it is for assessing similarities between countries. It merely is a practical problem in that it requires a very large number of comparisons to tease apart all the different criteria along which perceptual qualities can differ.

A further objection against similarity-based groupings is that similarity judgments are not stable over time. Instead, similarity judgments can be altered by experiences. In some cases, for example, odors can take on perceptual qualities of other odors that they have been experienced with previously (Stevenson 2001). Presumably, this change in perceived quality results in changes in similarity judgments of the odors involved. If we continuously experience vanilla odor and almond odor together, we tend to consider the two odors to be more similar than otherwise. It is certainly true that similarity judgments depend to some degree on experience. However, these experience-dependent effects on similarity judgments are minute compared to the effects of innate standards of similarity.[1] Although similarity judgments for very dissimilar odors change slightly depending on what other odors they were paired with, it seems unlikely that one could get a subject to rate fir oil as smelling more like fish than like pine oil by repeatedly exposing the subject to a mix of fir oil and fish. There are strong innate mechanisms of similarity perception that are unchanging. Furthermore, experience dependence of performance is common in all types of psychophysical testing. The capacity to discriminate odors, for example, is also affected by experience (Rabin 1988; Jehl et al. 1995).

In summary, for constructing a perceptual space that reflects similarity relations, similarity-based methods are the obvious choice. However, psychophysical similarity judgments can be problematic because they are cognitively more demanding than discrimination tasks. Humans are, for example, not good at comparing similarities between very dissimilar stimuli. Discrimination tasks are more stable and reliable and they have been the method of choice for constructing perceptual quality spaces. My goal is to construct an exhaustive perceptual quality space that includes the perceptual qualities of all modalities arranged by similarity relations between them. I do not know how this could be accomplished using discrimination tasks. I therefore propose to give up the methodological advantages of discrimination tasks to gain the possibility of constructing a perceptual space that arranges perceptual qualities according to their similarities regardless of their modality. I think it is worthwhile to sacrifice some precision in exchange for an exhaustive account of perceptual qualities. However, one can disagree. If there were no interesting similarity relations between stimuli in different modalities, then nothing is gained by the move from discrimination tasks to similarity judgments, and there would be no reason to attempt the construction of an exhaustive perceptual space. In the next section, I will speculate about features of the exhaustive perceptual space, in an attempt to show that interesting things about the relations between perceptual qualities in different modalities could be learned from such a space.

1.2 Dimensions of the Perceptual Quality Space

Using triadic comparisons to construct an exhaustive perceptual space is an ambitious empirical project. It would take a large number of triadic comparisons to reveal the detailed structure of this space. However, we already know enough about perceptual qualities and their similarity relations that one can speculate about the most important features of the perceptual quality space. In this section, I will speculate about the number of dimensions of the exhaustive perceptual space and about what the dimensions represent.

How Many Dimensions Does the Perceptual Quality Space Have?

An important question about the perceptual space is how many dimensions it has. The number of dimensions is a direct result of the similarity relations (Clark 1993, pp. 84–89). For every set of similarity relations, the lowest-dimensional space in which all of them can be represented can be found. Figure 1.2 illustrates how arranging 13 stimuli can require one, two, or three dimensions, based on the empirically determined discriminability of the stimuli pairs.

To illustrate how the dimensionality of the perceptual space is revealed by the similarity relations between perceptual qualities, consider sets of qualities that are all equally similar to one another. Consider three perceptual qualities: A, B, and C. If A is as similar to B as it is to C and as B is to C, then these three qualities cannot be arranged in a one-dimensional space in a way that reflects all three similarity relations. The simplest way to arrange three equidistant perceptual qualities in space is to place them in the corners of an equilateral triangle. Triangles are two-dimensional structures and the existence of three equidistant perceptual qualities shows that the perceptual space they belong to has to be at least two-dimensional. The simplest way to arrange four equidistant perceptual qualities is to place them into the four vertices of a regular tetrahedron, which is the only possible arrangement of four equidistant points in three-dimensional space. This means that a perceptual quality space that contains four equidistant perceptual qualities has to be at least three-dimensional.

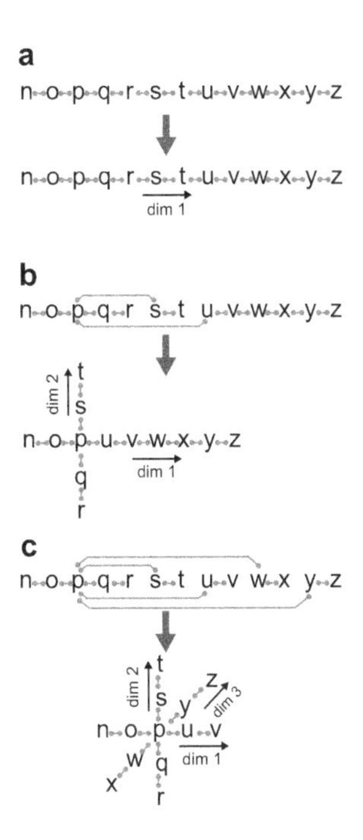

Fig.1.2 Determining the number of dimensions of a perceptual space. Depending on the experimentally determined discriminability of stimuli, one (**a**), two (**b**), or three (**c**) dimensions are required to arrange the 13 stimuli in perceptual space (example after [Clark 1993]). Indistinguishable perceptual qualities are connected by lines. In the *upper row* of each panel, the discriminability relations between the stimuli are shown. In the *lower row*, the lowest-dimensional spaces that can accommodate those relations are shown

We do not know yet how many dimensions are required to accommodate all perceptual qualities. However, partial quality spaces in which perceptual qualities from a single modality are arranged have been constructed. These quality spaces illustrate the explanatory power of this approach by arranging large numbers of perceptual qualities in low-dimensional spaces. The millions of colors can be arranged in a space that has three dimensions (hue, saturation, and brightness) (Hardin 1988; Hilbert and Kalderon 2000). The approximately 340,000 tones can be arranged in a two-dimensional space (loudness and pitch) (Stevens and Davis 1938). Temperature space and pressure space presumably only have a single dimension. There are many more smells than colors and tones (Bushdid et al. 2014). Accordingly, no smell space has yet been constructed. However, it has been speculated that an olfactory perceptual space would have many more dimensions than the quality spaces that we know now (Auffarth 2013).

Having an X-dimensional perceptual quality space does not mean that each perceptual quality in the space can only be fully described by X dimensions. It is possible that some perceptual qualities have zero values for many of the dimensions. Just because an X-dimensional space is required to accommodate *all* perceptual qualities does not mean that the X-dimensional space is required to accommodate *each* perceptual quality (Fig. 1.3). Grays are colors that differ only in their brightness. However, although the grays could be arranged in a one-dimensional quality space, they are commonly included in the three-dimensional space that can accommodate also those colors that have hue and saturation. Let us assume that the exhaustive perceptual space has ten dimensions. This does not mean that color perceptual qualities gain seven dimensions in addition to hue, brightness, and saturation.

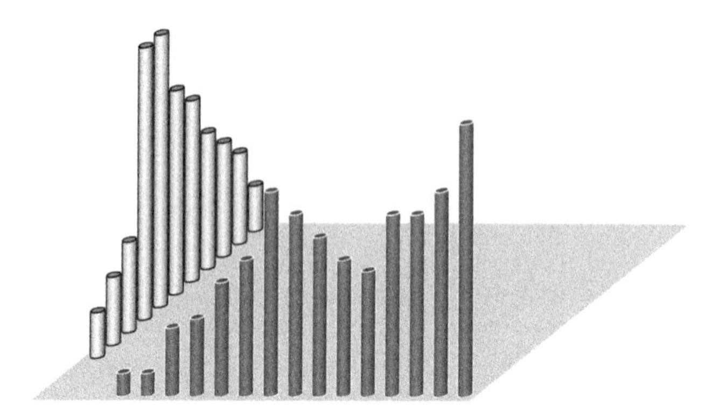

Fig. 1.3 Not all dimensions in a perceptual space are shared by all perceptual qualities. A hypothetical distribution of perceptual qualities in a three-dimensional quality space is shown. The perceptual qualities form two clusters, shown in *light gray* and *dark gray*. Perceptual qualities in both modalities vary along the y-axis (*bottom* to *top*), but only the perceptual qualities in the *dark gray* cluster vary along the x-axis (*left* to *right*) and only perceptual qualities in the *light gray* cluster vary along the z-axis (*front* to *back*)

What Are the Dimensions of the Perceptual Quality Space?

Once we know how many dimensions the perceptual space has, we will want to know what the dimensions represent; we will want to label the dimensions. Ultimately, it is a matter of convention how to label the dimensions of a perceptual space. The dimensions of the perceptual space may align with words that are used in pre-theoretic vocabulary to talk about perceptual qualities. However, it is also possible that the newly discovered dimensions of the perceptual space require new labels. Consider the labels for the dimensions of the color or tone space. "Hue" was an obscure Old English word for color that was revived by scientists in the nineteenth century to have a specific word for the perceptual property that is distinct from "color". In German, "hue" is "Farbton", which translates to "tone of the color", in analogy to auditory perception. "Pitch" was first used to describe perceptual auditory qualities in 1590, and the relations to the other meanings of the word are unclear. In German, "pitch" is "Tonhöhe", which translates to "height of the tone", reflecting the spatial position of tones in musical annotation. These examples show that labels for the dimensions of perceptual quality spaces were invented or re-appropriated by experts after the structure of the underlying perceptual space was understood. The same approach should be followed for labeling the dimensions of the exhaustive perceptual space.

That scientific progress brings with it a new vocabulary instead of relying on pre-theoretical language is a common occurrence. One example in which bringing systematic order to biological diversity required a new vocabulary is phylogenetic systematics and the naming of animal species. The pre-scientific vocabulary to refer to animals was not systematic. The many species of flying insects that bite or sting are called "mosquitoes". Myriads of small invertebrate species are referred to as "bugs". Only species that are of ecological importance to humans, like domesticated species or important pests, have common names. In some cases, cultivars or varieties also have proper names. With the rise of the biological sciences, a second vocabulary for talking about living things has been developed. Scientific names of species are based on the structure of the phylogenetic tree.

The new names for the species were invented by scientists. This scientific vocabulary was developed without any regard as to whether it aligns well with the established ways of referring to living things. The scientific vocabulary had little impact on language. We still say that we have been bitten by a mosquito, not that we have been bitten by *Anopheles gambiae*. However, the scientific vocabulary has the advantage that it reflects the empirically determined structure of the phylogenetic tree.

New scientific discoveries commonly require a new vocabulary. Otherwise, there would be no need for experiments and one could deduce the structure of reality from language use. The lack of pre-theoretical verbal labels for the dimensions of the perceptual space is not a surprise and does not pose a problem because our vocabulary has "no more of a *representational* relation to an intrinsic nature of things than does the anteater's snout or the bowerbird's skill at weaving" (Rorty 1998, p. 48). Contrary to this insight, a strong preference for an account of perception that aligns with the pre-theoretical vocabulary has also been expressed:

> It is therefore crucial to my thesis to emphasize that sense impressions or raw feels are common sense theoretical constructs introduced to explain the occurrence, *not* of white rat type discriminative behavior, but rather of perceptual propositional attitudes, and are therefore bound up with the explanation of why human language contains families of predicates having the logical properties of words for perceptible qualities and relations. (Sellars 1965)

My preferred strategy is to first construct the perceptual quality space and then attempt to describe it using non-scientific nomenclature. If this turns out to be impossible, a scientific vocabulary has to be invented. Admittedly, an account that would explain both perception *and* the way we talk about perception would be more powerful than an account of perception alone. However, I do not believe that such an account exists. Why we have words for some parts of the perceptual space and how these words relate to each other is a linguistic question. The structure of perceptual quality spaces, which can be constructed for bats or bacteria in the same manner they are constructed for humans, is independent of this linguistic question (for interesting treatments of the language used to describe perceptual qualities, see Berlin and Kay 1969 for colors, and Castro et al. 2013 for smells).

Even though building a perceptual space proceeds without regard of pre-theoretic language, it is likely that at least some of the dimensions of the perceptual space can be interpreted in pre-theoretical language. One of the dimensions of the exhaustive perceptual space will presumably represent perceptual intensity. Intensity is a quality shared by perception in many modalities, and stimuli of similar intensity in different modalities are perceived to be more similar than stimuli of different intensity. It has, for example, been shown that darker colors are matched to stronger odors (Kemp and Gilbert 1997). It is also likely that stronger odors will be matched to louder sounds and stronger pains. These results show that intensity is an important feature of perceptual qualities across modalities. This will likely be reflected by the structure of the perceptual space. A separate dimension of the perceptual space may correlate with perceived pleasantness, which is also a quality of perception that is found in different modalities. Pleasantness, for example, plays an important role in the non-random matching of odors to sounds (Crisinel and Spence 2011). For tastes and smells, pleasantness may even be the dominant perceptual dimension (Khan et al. 2007).[2]

Beyond intensity and pleasantness, correspondences between perceptual qualities in different modalities are more difficult to interpret using pre-theoretical language. Some crossmodal correspondences are easily explained through associations. The color green is matched to the smell of grass because the color and the smell are properties of the same object (Levitan et al. 2014). Other crossmodal correspondences, for example, between pitch and smell, cannot be explained through associations and are grounded in perceptual similarities across modalities (for a review of crossmodal correspondences between smells and perceptual qualities in other modalities, see Deroy et al. 2013). However, ordinary language has no word for a perceptual quality dimension that is shared by pitch and smell that could be used to label this emerging dimension of the perceptual quality space. A new label has to be invented for this dimension and presumably also for many other dimensions.

1.3 Modality-Representing Clusters of Perceptual Qualities

Many methods of bringing order to perceptual qualities start by assigning each perceptual quality to one of the sensory modalities.[3] The method suggested here arranges all perceptual qualities according to their similarities. The modalities, rather than being categories used to arrange perceptual qualities, emerge from the arrangement of perceptual qualities. In this section, I will discuss this method of individuating modalities based on the distribution of perceptual qualities in the exhaustive perceptual quality space.

Modality Individuation

The coordinate system that contains the exhaustive perceptual space is not fully packed with perceptual qualities. Instead, it is sparsely populated by perceptual qualities that are interspersed by empty areas that do not correspond to any perceptual qualities. Presumably, there is no continuum of perceptual qualities between a color of a certain hue, brightness, and saturation, and a tone of a certain pitch and loudness. We cannot start with a perceptual color quality and change it in small steps until we have changed it into a perceptual tone quality. If this is correct, then the area between colors and smells in the coordinate system that contains the exhaustive perceptual space is empty. Presumably, there will be many other empty areas surrounding clusters of perceptual qualities. These clusters of perceptual qualities can be interpreted as modalities. If people perceive colors to be more similar to other colors than to any other perceptual quality, then all colors will be close to each other in a cluster that can be interpreted as representing the visual modality.

Modality-representing clusters of perceptual qualities emerge when perceptual qualities are arranged according to their similarities. However, because a variety of different formalized algorithms (Everitt et al. 2011) can be employed to detect clusters, cluster analysis allows for several different outcomes. The same arrangement can be interpreted as three modalities (Fig. 1.4a), four modalities (Fig. 1.4b), or three modalities with one of the modality-representing clusters subdivided into subclusters (Fig. 1.4c). The hierarchical organization of modalities that is shown in Fig. 1.4c is

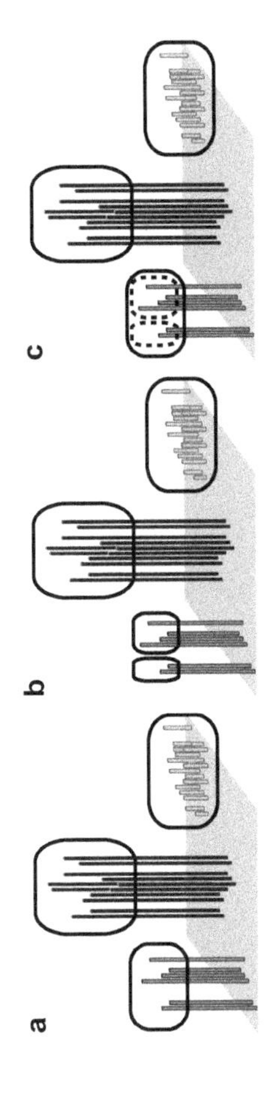

Fig. 1.4 Modality individuating clusters of perceptual qualities. A hypothetical distribution of perceptual qualities in a three-dimensional quality space is shown. Depending on the parameters that one uses for identifying modality-representing clusters, this distribution can be interpreted as three (**a**) or four (**b**) modalities, or as three modalities, one of which can be further subdivided into two subclusters (**c**)

probably the most accurate reflection of the relations between perceptual qualities. The best-known illustration of such a hierarchical structure is the arrangement of the perceptual qualities associated with touch. Touch includes every perception that is not vision, audition, gustation, or olfaction. Touch can therefore be further subdivided into proprioception (sensing relative positions of body parts), mechanoception (pressure), thermoception (temperature), nociception (pain), and maybe others.

Because the results of applying cluster analysis to the perceptual space depend on the criteria used to individuate clusters, cluster analysis cannot decide unambiguously how many modalities there are and where the boundaries between them are located. However, the advantage of cluster analysis is that it can be used to test accounts of modalities for their consistency. For example, the set of algorithms that clusters smells, tastes, colors, and tones all in their own modality-representing cluster can be investigated. If no algorithm in this set clusters all other perceptual qualities together in a single touch modality, then the classic account of the five human senses is wrong.

The Relation Between Modalities

The big advantage of modality individuation through modality-representing clusters over other strategies of modality individuation is that it not only individuates modalities but also reveals the relation between modalities. The distance between two modality-representing clusters in the perceptual space is a measure of the similarity between the two modalities. More interestingly, the cluster analysis also reveals the relative orientation of modality-representing clusters to each other. The distance between these clusters and how they are oriented with respect to each other represent similarity relations between perceptual qualities in different modalities. Consider again that sugar, an odorless tastant, is consistently experienced to be more similar to the tasteless odorant vanillin than to salt, another odorless tastant (Rankin and Marks 2000). The traditional two-step procedure of modality individuation and subsequent quality space construction cannot account for this fact. In contrast, the perceptual quality space that contains all perceptual qualities can account for all similarity relations between perceptual qualities, regardless of the modality-representing cluster they belong to. In Fig. 1.5, a hypothetical arrangement of the cluster of gustatory perceptual qualities and olfactory perceptual qualities is shown. This arrangement preserves the distinction between the olfactory modality and the gustatory modality while also explaining why sweet taste is more similar to vanilla odor than to salty taste.

Interpreting clusters of perceptual qualities in the perceptual space as modalities is a form of individuating modalities based on behaviors, because the perceptual space was constructed through behavioral experiments. This is an unusual strategy. It is more common that modality individuation is based on representations, phenomenal character, the proximal stimulus, or the sense organs (Grice 1962; Macpherson 2011).[4] However, behavior-based modality individuation has also been suggested previously. Most notably, the just-noticeable-difference method that has been used to construct quality spaces has recently been extended to provide a method for individuating modalities based on discriminative abilities (Rosenthal 2014). The just-noticeable-difference method and the method proposed in Sect. 1.1 have many similarities. The most notable difference between the two methods is that the method proposed in this chapter arranges perceptual modalities relative to each other according to the similarities between their perceptual qualities. As was

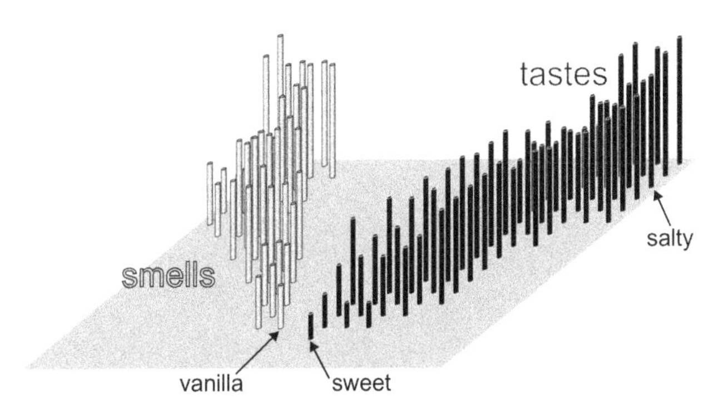

Fig. 1.5 Similarity between perceptual qualities in different modalities. A hypothetical arrangement of smells and tastes in perceptual space that would explain why sweet taste is more similar to vanilla odor than to salty taste despite sweet and salty both being tastes, is shown

pointed out above, the discriminative abilities that are revealed by the just-noticeable-difference method are also revealed by relative similarity judgments. When subjects are presented with three stimuli, two of which are identical and the third different, triadic comparisons mimic discrimination tasks. It can therefore be suspected that the modality individuation through both methods results in the same set of modalities. The method proposed in Sect. 1.1 individuates the modalities in a similar way to the just-noticeable-difference method and constructs similar perceptual spaces for each individual modality. However, it also arranges the modality-representing clusters of perceptual qualities in a way that reflects the similarities between perceptual qualities in different modalities.

1.4 Orderliness of the Perceptual Quality Space

The perceptual tone space and the perceptual color space have explanatory power because they arrange a large number of perceptual qualities in orderly spaces. The approximately 340,000 tones (Stevens and Davis 1938) can be arranged in a two-dimensional space (loudness and pitch) and the millions of discriminable colors (Nickerson and Newhall 1943; Pointer and Attridge 1998) can be arranged in three dimensions (hue, saturation, and brightness) (Hardin 1988; Hilbert and Kalderon 2000). This means that each of millions of colors can be identified by three coordinates that represent hue, saturation, and brightness. Giving a full account of perceived colors therefore does not require a description of millions of colors, but only a description of three dimensions.

The reduction in complexity from hundreds of thousands or even millions of perceptual qualities to two or three dimensions is what makes the color space and the tone space useful. They would lose their usefulness if they were less orderly and arrange perceptual qualities in spaces with hundreds instead of two or three dimensions. If it were not possible to arrange all perceptual qualities in a low-dimensional space, the exhaustive perceptual space would not contribute to our understanding of perception. The perceptual qualities that so far have been most resistant to being arranged in a perceptual space are smells. There have been so many failed attempts to construct a smell space that some doubt that it is possible (Lycan 2000). If there would be no orderly smell space, there also would be no orderly space that includes all perceptual qualities and the project outlined here would not succeed.

Orderliness in the Olfactory Perceptual Space

After reviewing the many failed attempts to put order to the olfactory perceptual space, Lycan writes, "smell may catch up to color and hearing in the orderliness of its quality space and in the accompanying explanatoriness of the underlying neurophysiology. But that is not *overwhelmingly* likely" (Lycan 2000, p. 280). Such pessimism is based on the failure of all previous attempts to arrange smells in a perceptual space (Berglund and Höglund 2012; Kaeppler and Mueller 2013). However, these failures could also be due to three main shortcomings in the attempts to build smell spaces.

One common mistake is that many approaches are not driven by perceptual qualities but by verbal labels for smells. Researchers often take odor labels like "burnt", "fragrant", and "spicy", and then arrange olfactory perceptual qualities according to how burnt, fragrant, and spicy they smell. An example that is still found in psychology textbooks is Henning's smell prism (Henning 1916), which is a triangular prism, the six corners of which are labeled "flowery", "foul", "fruity", "spicy", "burnt", and "resinous". All olfactory perceptual qualities are supposed to be located within this three-dimensional space. Projects like this conflate constructing a perceptual smell space with establishing smell categories. The dimensions of the three-dimensional perceptual color space are hue, saturation, and brightness. This space could have never been discovered using an approach in which color categories like "red", "green", and "blue" are taken to be foundational. On closer examination, it turns out that most of the projects that seem like failed attempts to establish a smell space were instead successful attempts to establish an "odor descriptor space" (Castro et al. 2013). The attempts were successful in establishing the odor descriptor space. However, the odor descriptor space does not correspond to a perceptual space. This lack of correspondence is to be expected since it is also seen with the color space that the linguistic analysis of color terms does not result in a better understanding of the structure of the perceptual color space.

The second problem faced by researchers trying to elucidate the structure of the smell space is the difficulty of finding the right odorants to use for such a project. The number and diversity of odorous molecules are immense. More than one billion different chemicals have a smell (Joel Mainland, personal communication). Choosing a few dozens of these stimuli to use in psychophysical experiments is a daunting task. Traditionally, there was no attempt to find representative stimuli for psychophysics. Instead, especially salient, interesting, and ecologically or economically important stimuli have been used. Avery Gilbert and Mark Greenberg succinctly summarized the danger of this predominant approach when they wrote that "we are creating a science of olfaction based on cinnamon and coffee" (Gilbert and Greenberg 1992, p. 329). This is equivalent to a color science based on the colors of earth, fire, and water. It is not surprising that attempts to construct smell spaces based on perfumery raw materials failed. Thanks to the work of Noam Sobel and his group, now sets of odorants that are representative of all odorants can be identified. Sobel and coworkers accomplished this by arranging odorous chemicals according to the similarity of their chemical and physical properties in a stimulus space. Once such a stimulus space has been established, sets of odorants that are representative for this stimulus space can be identified (Haddad et al. 2008). It is therefore, for the first time, possible to use representative odorants in psychophysical experiments and then generalize the results of these experiments to all of olfaction.

The third shortcoming of most attempts to construct a smell space is that they are based on individual odorous molecules (benzaldehyde, hexanal, vanillin, etc.) (Wise et al. 2000). A perceptual space that is based on the smells of individual molecules only covers a tiny fraction of all olfactory qualities because mixtures of odorous molecules frequently have qualities that are different from the qualities of its components and of all other odorous molecules. To construct an olfactory perceptual space that covers all olfactory perceptual qualities, the olfactory qualities of odor mixtures have to be included. The perceptual qualities of odor mixtures are the qualities we are familiar with because the smells encountered in nature are usually mixtures. The characteristic scents of a rose (Ohloff 1994, pp. 154–158), coffee (Grosch 1998), and red wine (Aznar et al. 2001) are all complex mixtures of hundreds of components. The components of these mixtures interact in a variety of ways that are not well understood (Ferreira 2012a, b).

Previous attempts to reveal the structure of the smell space have failed. However, concluding from these failures that no orderly smell space exists would be premature. Before coming to this conclusion, it is worthwhile to try an approach that does not rely on our limited olfactory vocabulary and uses mixtures of representative odorants. Such a semantic-free approach, which is more similar to the approaches that were successfully applied to colors and tones, may result in the discovery of an orderly smell space.

Size of the Olfactory Perceptual Space

How orderly the space that contains all perceptual qualities is depends on how orderly the smell space is. Most perceptual qualities that we can discriminate are olfactory perceptual qualities. There are more than one trillion discriminable olfactory stimuli (Bushdid et al. 2014). The actual number is higher because the estimate of one trillion is based on mixtures of 30 out of a collection of 126 different odorous molecules. However, many more than 126 different odorous molecules exist and molecules can be mixed into mixtures of more than 30 components. Furthermore, the odors can be mixed at different proportions. Whatever the actual number of olfactory qualities is, it is several orders of magnitude higher than the number of visual and auditory qualities.

A human takes less than one billion breaths in his or her lifetime. We encounter only relatively few ecologically relevant odors regularly. It is therefore surprising that we can discriminate so many different smells. We did not evolve to be able to distinguish such a large number of smells. Instead, we evolved to be able to discriminate minute differences between two olfactory stimuli. The large number of discriminable smells is a consequence of our evolved discriminatory capacity. Olfactory systems evolve to discriminate complex mixtures that differ in few of their components. In pre-historic times, it may have been important to distinguish the smell of several babies that were raised together to avoid feeding unrelated babies in times of scarcity. The body odors of related babies that live together are very similar. Body odors are complex mixtures that share many components but also differ in a few. Similarly, the odor of food and the same food that shows the first signs of being spoiled can be very similar. Again, the two different stimuli have many shared components, but they also differ in the components that are produced by the bacteria as the food starts to spoil. Detecting this difference can mean the difference between a nutritious meal and food poisoning. The ease with which we detect cork taint is a good example of the incredible resolution of our olfactory system. Detecting cork taint amounts to detecting the presence of a small amount of the chemical 2,4,6-trichloroanisole in the mixture of hundreds of diverse molecules that make up the wine aroma (Aznar et al. 2001).

Because minute amounts of chemicals that are added to complex mixtures of other chemicals usually lead to a change in the perceived smell of the mixture, and because the number of potentially odorous chemicals is in the billions, it has been suggested that the number of distinguishable olfactory stimuli is "unlimited" (Wright 1964, p. 80). If the number of perceptual qualities were indeed unlimited, no perceptual quality space that contains all of them could be constructed. However, a recent breakthrough in olfactory psychophysics showed that our ability to discriminate smells is limited. Noam Sobel and colleagues showed that mixtures with many components, when each component is diluted so that all components have similar intensity, converge perceptually. This means that mixtures of random odorous molecules with a large enough number of components smell similar and share an olfactory quality that has been called, in analogy to vision, "olfactory brown" or "olfactory white" (Weiss et al. 2012). The reason why the complex mixtures of odorous molecules that we encounter when we smell roses or coffee do not smell similar is that the components of these mixtures are not a random sampling of odorous molecules and the molecules do not contribute equally to the smell of the mixture. Instead, in many cases, the smell of a natural mixture is dominated by a few components. How many components are necessary in mixtures to render them indistinguishable from one another is not yet known (Weiss et al. 2012). However, that larger mixtures converge perceptually shows that the resolution of the olfactory system is limited. The number of perceptual qualities is finite and they therefore can be arranged in a finite perceptual quality space.

There are two possibilities of how a perceptual space can accommodate a very large number of perceptual qualities. A large number of perceptual qualities can either result from a high resolution along the dimensions of the quality space or from a large number of dimensions. Quality spaces are mathematical constructs that have whatever number of dimensions is needed to capture the similarity relations of the relevant qualities (see Fig. 1.2). It is unlikely that the odor space will have only two or three dimensions like the tone or color space. Smell space has been suggested to have a higher dimensionality than quality spaces in other modalities (Berglund and Höglund 2012; Auffarth 2013). It simply is not possible to arrange all olfactory qualities in a low-dimensional space. However, it is also unlikely that the number of dimensions of the smell space is very high. The number of perceptual qualities that can be accommodated in a perceptual space increases exponentially with the number of dimensions of the space. Maybe the smell space will reduce the complexity by a similar degree as perceptual spaces in other modalities. Approximately five million discriminable colors are arranged in a three-dimensional space and around 340,000 tones are arranged in a two-dimensional space. The cube root of 5,000,000 is 171 and the square root of 500,000 is 583, which means that each dimension of these spaces can be divided into a few hundred discrimination steps. If in the olfactory space 500 discriminable perceptual qualities are arranged along each dimension, then five dimensions would be required to accommodate a trillion olfactory perceptual qualities.

How orderly the exhaustive perceptual space is, is an empirical question that can only be answered with certainty after the psychophysical experiments required to construct the exhaustive perceptual space have been performed. Partial perceptual spaces are already known. The color and the tone space arrange a large number of perceptual qualities in spaces with three and two dimensions. It is likely that spaces for temperature or touch qualities will also have few dimensions. The only potential problem is how to arrange smell qualities. The smell space is likely to have more dimensions than the color space or the tone space. However, although we will have to await the construction of the smell space to be sure, I am optimistic that the smell space will not have more than hundreds of dimensions. It is likely that the smell space will increase order by a similar degree as the perceptual spaces in the other modalities.[5] There is hope that it will reduce complexity enough to be a useful construct for perceptual philosophy. If this is the case, then the exhaustive perceptual space will be orderly enough to play an important role in our understanding of how perception works.

1.5 Conclusion: Perceptual Qualities Can Be Individuated by Their Position in a Similarity Space

All perceptual qualities perceived by a perceiver can be individuated based on their similarity relations with other perceptual qualities. Sensory modalities can then be interpreted as clusters of perceptual qualities in this exhaustive perceptual space. The relational account of perceptual qualities that I developed here is based on the Quality Space Theory (Rosenthal 2005, 2010), but it uses different behavioral methods and therefore can account for similarity relations between perceptual qualities in different modalities. Relational accounts of perceptual qualities are alternatives to accounts that determine perceptual qualities by their intrinsic properties. Accounts that individuate perceptual qualities by intrinsic properties often lead to the conclusion that the only thing that can provide access to perceptual qualities is first-hand conscious experience. A relational account avoids this conclusion by showing how we can describe every perceiver's perceptual qualities.

Importantly, perceptual qualities are arranged in the perceptual space based on behavioral experiments and regardless of whether they are consciously perceived or not. What it feels like to perceive a certain perceptual quality is not used as a basis to construct the perceptual space. Instead, what it feels like to perceive a certain perceptual quality reflects its position in the perceptual space. The perceptual space is a mechanistic explanation of the qualities of subjective awareness. Individual instances of subjective awareness are not necessarily relational. At any time, we just smell a single odor. There are no relations to other perceptual olfactory qualities. However, how the odor smells is determined by its position in the perceptual space. This position can be expressed by coordinates along the dimensions of the perceptual space.

Notes

1. As Quine pointed out, if all stimuli would be equally alike and equally different, we could never acquire generalized behavioral responses. Quine (1970). Natural kinds. *Essays in Honor of Carl G. Hempel*. N. Rescher. Dordrecht, D. Reidel Publishing Company. All our behavioral responses are generalized responses that are elicited by a group of similar yet discriminable stimuli.
2. There is a question whether pleasantness should count as a dimension of perceptual qualities. The question mirrors the question whether painfulness should count as a feature of perceptual qualities, which has been discussed by Austen Clark (2005). Painfulness is Not a Quale. *Pain: New Essays on Its Nature and the Methodology of Its Study*. M. Aydede. Cambridge, MIT Press: 177–197. According to Clark, pain is accompanied by specific sensory qualities, but it is not one itself. The intuitive test whether one agrees with this is whether two equally painful pains, for example, the pain of a sunburn and of a pulled muscle, share a common sensory quality or not. The olfactory equivalent would be to ask whether two equally pleasant smells, for example, the smell of bacon and that of vanilla, share a common quality or not. Clark concludes that pain is not a perceptual quality, instead, the close connection between certain perceptual qualities and pain is that the sensory qualities are wired "directly into the creature's preference functions" ibid. Whether something that is directly wired to perception is part of perception or not is part of the difficult problem of drawing the line between perception and cognition, which I will discuss in Chapter 6.
3. A notable exception is Rosenthal, who uses the just-noticeable-difference method to both construct perceptual quality spaces and individuate modalities. Rosenthal (2014). Quality spaces and sensory modalities. *The Nature of Phenomenal Qualities: Sense, Perception, and Consciousness*. P. Coates and S. Coleman. Oxford, Oxford University Press.
4. How the sense modalities should be individuated has been a topic of philosophical investigations since Aristotle's *De Anima*. Aristotle (2011). Excerpt from On the Soul (De Anima). *The Senses: Classical and Contemporary Philosophical Perspectives*. F. Macpherson. New York, Oxford University Press: 47–63; Sorabji (2011). Aristotle on Demarcating the Five Senses. *The Senses: Classical and Contemporary Perspectives*. F. Macpherson. New York, Oxford University Press: 64–82. Traditionally, modality individuation has been based on representations, phenomenal character, the proximal stimulus, or the sense organs. All four approaches

largely agree when individuating vision and audition. Vision, for example, is the only sense that represents colors. Our eyes are at all levels of description (anatomical, cellular, and molecular) sufficiently distinct from other sensory structures. Photons, the proximal stimulus, are also sufficiently distinct from other stimuli. The phenomenal character of all visual experiences allows distinguishing these experiences from other sensory experiences. However, it is merely a contingent fact of the biology of human vision that all four approaches agree on how to individuate it. With olfaction and other modalities they often produce contradictory results, as has been anticipated by Aristotle, who wrote "the distinguishing characteristic of smell is less obvious than those of sound or colour" Aristotle (2011). Excerpt from On the Soul (De Anima). *The Senses: Classical and Contemporary Philosophical Perspectives*. F. Macpherson. New York, Oxford University Press: 47–63. (page 52). One example of chemosensory perception that is difficult to categorize using the traditional method is the perception mediated by the TRPV1 receptor. TRPV1 is a type of molecular receptor that is sensitive both to hot temperature and to capsaicin, the pungent chemical found in chili peppers. Caterina et al. (1997). "The capsaicin receptor: a heat-activated ion channel in the pain pathway." *Nature* **389**(6653): 816–824. This receptor is expressed in sensory neurons on the tongue that respond equally to chili peppers and to hot water. If the sense-organ criterion is applied, capsaicin and heat are considered two stimuli in the same modality. However, if the stimulus criterion is applied, then the TRPV1 receptor mediates perception in two different modalities. Two stimuli that are sensed by the same molecular receptor will result in the same neuronal activity and therefore in the same phenomenal character; therefore, the phenomenal character criterion would judge heat and capsaicin to be two stimuli in the same modality. However, what is represented by the two stimuli is a botanical compound and hot air, respectively. TRPV1 is just one of many examples of chemical receptors that are sensitive to physically different stimuli Dhaka et al. (2006). "Trp ion channels and temperature sensation." *Annual Review of Neuroscience* **29**: 135–161. Another prominent example is the TRPM8 receptor, which is activated by the molecule menthol as well as by cold air. In these cases of receptors that are sensitive to two physically very different stimuli, the stimulus and representation criteria lead to the conclusion that the receptor mediates perception in two different modalities whereas the phenomenal character and sensory organ criteria come to the conclusion that the two different stimuli are stimuli in the same modality.

5. The procedure for constructing perceptual spaces described in Sect. 1.1. will result in a perceptual space regardless of the similarity relations between the perceptual qualities. Any set of perceptual qualities can be arranged in a multidimensional space according to their similarity. It is therefore not possible that there is no perceptual smell space. However, a perceptual space in which a trillion perceptual qualities are arranged in a trillion-dimensional perceptual space would be nothing more than a list of the perceptual qualities and not a useful theoretical construct.

References

Aristotle. (2011). Excerpt from On the Soul (De Anima). In F. Macpherson (Ed.), *The senses: Classical and contemporary philosophical perspectives* (pp. 47–63). New York: Oxford University Press.

Auffarth, B. (2013). Understanding smell: The olfactory stimulus problem. *Neuroscience and Biobehavioral Reviews, 37*(8), 1667–1679.

Aznar, M., López, R., et al. (2001). Identification and quantification of impact odorants of aged red wines from Rioja. GC-olfactometry, quantitative GC-MS, and odor evaluation of HPLC fractions. *Journal of Agricultural and Food Chemistry, 49*(6), 2924–2929.

Berglund, B., & Höglund, A. (2012). Is there a measurement system for odour quality? In G. M. Zucco, R. S. Herz, & B. Schaal (Eds.), *Olfactory cognition: From perception and memory to environmental odours and neuroscience* (pp. 3–21). Amsterdam: John Benjamins Publishing Company.

Berlin, B., & Kay, P. (1969). *Basic color terms: Their universality and evolution.* Berkeley: University of California Press.

Bushdid, C., Magnasco, M. O., et al. (2014). Humans can discriminate more than one trillion olfactory stimuli. *Science, 343*(6177), 1370–1372.

Castro, J. B., Ramanathan, A., et al. (2013). Categorical dimensions of human odor descriptor space revealed by non-negative matrix factorization. *PLoS One, 8*(9), e73289.

Caterina, M. J., Schumacher, M. A., et al. (1997). The capsaicin receptor: A heat-activated ion channel in the pain pathway. *Nature, 389*(6653), 816–824.

Chen, Y. C., & Gerber, B. (2014). Generalization and discrimination tasks yield concordant measures of perceived distance between odours and their binary

mixtures in larval Drosophila. *Journal of Experimental Biology, 217*(12), 2071–2077.
Churchland, P. (2005). Chimerical colors: Some phenomenological predictions from cognitive neuroscience. *Philosophical Psychology, 18*(5), 527–560.
Clark, A. (1993). *Sensory qualities*. Cambridge: Cambridge University Press.
Clark, A. (2005). Painfulness is not a quale. In M. Aydede (Ed.), *Pain: New essays on its nature and the methodology of its study* (pp. 177–197). Cambridge: MIT Press.
Crisinel, A. S., & Spence, C. (2011). A fruity note: Crossmodal associations between odors and musical notes. *Chemical Senses, 37*(2), 151–158.
Deroy, O., Crisinel, A.-S., et al. (2013). Crossmodal correspondences between odors and contingent features: Odors, musical notes, and geometrical shapes. *Psychonomic Bulletin & Review, 20*(5), 878–896.
Dhaka, A., Viswanath, V., et al. (2006). Trp ion channels and temperature sensation. *Annual Review of Neuroscience, 29*, 135–161.
Everitt, B. S., Landau, S., et al. (2011). *Cluster analysis*. Hoboken: Wiley.
Ferreira, V. (2012a). Revisiting psychophysical work on the quantitative and qualitative odour properties of simple odour mixtures: A flavour chemistry view. Part 1: Intensity and detectability. A review. *Flavour and Fragrance Journal, 27*(2), 124–140.
Ferreira, V. (2012b). Revisiting psychophysical work on the quantitative and qualitative odour properties of simple odour mixtures: A flavour chemistry view. Part 2: Qualitative aspects. A review. *Flavour and Fragrance Journal, 27*(3), 201–215.
Gilbert, A. N., & Greenberg, M. S. (1992). Stimulus selection in the design and interpretation of olfactory studies. In M. J. Serby & K. L. Chobor (Eds.), *Science of olfaction* (pp. 309–334). New York: Springer-Verlag.
Goodman, N. (1951). *The structure of appearance*. Cambridge: Harvard University Press.
Gray, R. (2013). Is there a space of sensory modalities? *Erkenntnis, 78*, 1259–1273.
Grice, H. P. (1962). Some remarks about the senses. In R. J. Butler (Ed.), *Analytical philosophy* (Vol. 1). Oxford: Oxford University Press.
Grosch, W. (1998). Flavour of coffee. *Food/Nahrung, 42*(6), 344–350.
Haddad, R., Khan, R., et al. (2008). A metric for odorant comparison. *Nature Methods, 5*(5), 425–429.
Hardin, C. L. (1988). *Color for philosophers*. Indianapolis: Hackett Publishing Company.
Henning, H. (1916). *Der Geruch*. Leipzig: Barth.

Hilbert, D. R., & Kalderon, M. (2000). Color and the inverted spectrum. In S. Davis (Ed.), *Color perception: Philosophical, psychological, artistic, and computational perspectives*. Oxford: Oxford University Press.

Jehl, C., Royet, J. P., et al. (1995). Odor discrimination and recognition memory as a function of familiarization. *Perception and Psychophysics, 57*, 1002–1011.

Kaeppler, K., & Mueller, F. (2013). Odor classification: A review of factors influencing perception-based odor arrangements. *Chemical Senses, 38*(3), 189–209.

Keeley, B. L. (2002). Making sense of the senses: Individuating modalities in humans and other animals. *The Journal of Philosophy, 99*(1), 5–28.

Kemp, S. E., & Gilbert, A. N. (1997). Odor intensity and color lightness are correlated sensory dimensions. *The American Journal of Psychology, 110*(1), 35–46.

Khan, R. M., Luk, C. H., et al. (2007). Predicting odor pleasantness from odorant structure: Pleasantness as a reflection of the physical world. *Journal of Neuroscience, 27*(37), 10015–10023.

Levitan, C. A., Ren, J., et al. (2014). Cross-cultural color-odor associations. *PLoS One, 9*(7), e101651.

Lycan, W. G. (2000). The slighting of smell. In N. Bhushan & S. Rosenfeld (Eds.), *Of minds and molecules: New philosophical perspectives on chemistry* (pp. 273–289). Oxford: Oxford University Press.

Macpherson, F. (2011). Taxonomizing the senses. *Philosophical Studies, 153*, 123–142.

Macpherson, F. (2015). The space of sensory modalities. In D. Stokes, M. Matthen, & S. Biggs (Eds.), *Perception and its modalities* (pp. 432–461). Oxford: Oxford University Press.

MacRae, A. W., Howgate, P., et al. (1990). Assessing the similarity of odors by sorting and by triadic comparison. *Chemical Senses, 15*(6), 691–699.

MacRae, A. W., Rawcliffe, T., et al. (1992). Patterns of odor similarity among carbonyls and their mixtures. *Chemical Senses, 17*(2), 119–125.

Matthen, M. (2005). *Seeing, doing, and knowing*. Oxford: Oxford University Press.

Nickerson, D., & Newhall, S. M. (1943). A psychological color solid. *Journal of the Optical Society of America, 33*(7), 419–422.

Ohloff, G. (1994). *Scent and fragrances: The fascination of odors and their chemical perspectives*. Berlin: Springer-Verlag.

Pointer, M. R., & Attridge, G. G. (1998). The number of discernible colours. *Color Research and Application, 23*(1), 52–54.

Quine, W. V. (1970). Natural kinds. In N. Rescher (Ed.), *Essays in honor of Carl G. Hempel*. Dordrecht: D. Reidel Publishing Company.

Rabin, M. D. (1988). Experience facilitates olfactory quality discrimination. *Perception and Psychophysics, 44*, 532–540.
Rankin, K. M., & Marks, L. E. (2000). Chemosensory context effects: Role of perceived similarity and neural commonality. *Chemical Senses, 25*(6), 747–759.
Rorty, R. (1998). *Truth and progress: Philosophical papers* (Vol. 3). Cambridge: Cambridge University Press.
Rosenthal, D. M. (2005). *Consciousness and mind.* Oxford: Oxford University Press.
Rosenthal, D. M. (2010). How to think about mental qualities. *Philosophical Issues: Philosophy of Mind, 20*, 368–393.
Rosenthal, D. (2014). Quality spaces and sensory modalities. In P. Coates & S. Coleman (Eds.), *The nature of phenomenal qualities: Sense, perception, and consciousness*. Oxford: Oxford University Press.
Sellars, W. (1965). The identity approach to the mind-body problem. *Review of Metaphysics, 18*, 430–451.
Sorabji, R. (2011). Aristotle on demarcating the five senses. In F. Macpherson (Ed.), *The senses: Classical and contemporary perspectives* (pp. 64–82). New York: Oxford University Press.
Stevens, S. S., & Davis, H. (1938). *Hearing, its psychology and physiology.* New York: John Wiley.
Stevenson, R. J. (2001). Associative learning and odor quality perception: How sniffing an odor mixture can alter the smell of its parts. *Learning and Motivation, 32*(2), 154–177.
Weiss, T., Snitz, K., et al. (2012). Perceptual convergence of multi-component mixtures in olfaction implies an olfactory white. *Proceedings of the National Academy of Sciences, 109*(49), 19959–19964.
Wise, P. M., Olsson, M. J., et al. (2000). Quantification of odor quality. *Chemical Senses, 25*(4), 429–443.
Wright, R. H. (1964). *The science of smell.* London: George Allen & Unwin.
Young, B. D., Keller, A., et al. (2014). Quality space theory in olfaction. *Frontiers in Psychology, 5*, 1.

2

Third-Person Access to Perceptual Qualities

In Chap. 1, I have introduced a strategy to arrange all perceptual qualities according to the similarity relations between them. The resulting exhaustive perceptual space allows individuating each perceptual quality through its position within the space. This is an important step toward a science of perceptual qualities because it makes it possible to identify and compare perceptual qualities without having to refer to the subjective experiences of the perceiver.

However, the procedure suggested in the previous chapter does not result in a universal perceptual space, or in *the* human perceptual space, or even in a stable perceptual space of a given individual. This is not a shortcoming of the proposed strategy. There is neither a universal perceptual space, nor *the* human perceptual space, nor an individual's unchanging perceptual space. The experimental strategy introduced in Chap. 1 produces an exhaustive perceptual space for a given perceiver at a given time. If one is interested in another perceiver's perceptual space, the experiments have to be repeated with the other perceiver. The perceptual spaces of two perceivers will differ. It is therefore necessary to develop a strategy of comparing perceptual qualities in different perceptual spaces. If no such strategy is available, then third-person access to perceptual

A. Keller, *Philosophy of Olfactory Perception*,
DOI 10.1007/978-3-319-33645-9_2

qualities would not be possible and studying perceptual qualities in a systematic way would be a very daunting task.

In this chapter, I will present a strategy for comparing perceptual qualities that are perceived by different perceivers. The strategy that I suggest is to first construct the perceptual spaces of the two perceivers and then register them by transforming them into one coordinate system. To set up the problem, I will first review the diversity of perceptual spaces. In the second part of the chapter, I will then explain how other perceiver's perceptual qualities can be accessed under certain circumstances.

2.1 Diversity of Perceptual Spaces

Perceptual quality spaces can be constructed for every perceiver that can be made to respond differently to different stimuli. The perceivers can be humans, animals, plants, machines, robots, or extraterrestrials. Each perceiver has a different mechanism for perceiving and for processing sensory information. Different types of robots and machines are built according to different specifications. Different species evolved different perceptual systems and brains that are adapted to the perceptual needs of the species' ecological niche. Perceivers belonging to the same species can also vary considerably, and the brain and sense organs of a given perceiver change over time. This diversity of perceptual systems is reflected by the diversity of perceptual spaces.

The realization that there are as many perceptual spaces as there are perceivers could be discouraging and interpreted as a sign that perceptual spaces are not useful constructs for theorizing about perception. However, this would be the wrong response. The diversity and malleability of perceptual spaces does not reduce the explanatory power of these constructs. That lack of universality does not necessarily reduce explanatory power can be illustrated by the role of genomes in understanding phenotype determination. Like perceptual quality spaces, genomes are diverse. Different species have different genomes, and individuals of the same species carry different genetic variants of many genes in their genomes. Genomes also change over time due to mutations. Rather than diminishing the explanatory power of genomes, this diversity is the reason why genomes have become so important for our understanding of biology.

Differences in genomes explain anatomical and physiological differences between species. The differences between the genomes of members of the human species explain why some humans have blue eyes and why some are lactose intolerant. Changes in the genomes of our cells over time are the cause of cancer. That it is not possible to identify *the* genome, or *the* human genome, or *the* genome of a given individual does not diminish the important role that genomes play in understanding biology. Similarly, the diversity of perceptual spaces does not reduce their usefulness for our understanding of perception.

However, the diversity of perceptual spaces has to be acknowledged and the differences between perceptual spaces have to be known for perceptual spaces to be useful theoretical constructs. In this section, I will use the relatively well-understood color space as an example of how perceptual spaces differ between different perceivers. Perceptual spaces are arrangements of perceptual qualities according to similarities between them within a multidimensional coordinate system. Two perceptual spaces can have the same coordinate system but differ in the arrangement of perceptual qualities within that coordinate system. Alternatively, perceptual spaces can differ in the coordinate system in which the perceptual qualities are arranged.

Arrangements of Perceptual Qualities in Perceptual Spaces

In humans, the perceptual color quality space has three dimensions: hue, saturation, and brightness. In this section, I will consider how color spaces that have these three dimensions can differ from one another. In the next section, I will then speculate about color quality spaces with different dimensions.

The color spaces of different mammalian species differ considerably. Color perception in mammals is the consequence of differential activation of sensory neurons that carry different types of cone photopigments. These cone photopigments, like all proteins, are encoded by genes and these genes, called opsin genes, are subject to evolution through natural selection. In mammals, there have been many dramatic evolutionary changes in the opsin genes (Jacobs 2009). Opsin genes can change, which can alter their sensitivity to light of specific wavelengths. Additionally, opsin genes can duplicate and then change, which results in species that have more opsin genes than their ancestral species. Not only can species gain opsin genes, they can also lose them. An opsin gene is lost when a mutation that renders the photopigment non-functional occurs. Humans have three opsin genes; they are trichromatic. Most other mammals have only two opsin genes. They are dichromatic. Some species, like the owl monkey, are monochromats that have only a single opsin gene. On the other hand, some birds have more than three opsin genes. Most birds are tetrachromatic, while pigeons are pentachromatic. Behavioral experiments have shown that species with fewer opsin genes generally are worse at discriminating colored stimuli (for a review of comparative color vision, see Jacobs 1981). Additional opsin genes convey the ability to discriminate more color stimuli.[1] The distribution of color qualities in the hue-saturation-brightness-space is different depending on the number of types of receptors with which light is perceived.

In addition to these differences between different species, there are also substantial differences between individuals that belong to the same species, as can be illustrated by human variability in color vision. There are different forms of color blindness in humans. Some humans are dichromats. They lack one of the three cone photopigments. Depending on which of the three cone photopigments is missing, this condition is known as protanopia, deuteranopia, or tritanopia. In Fig. 2.1, the color qualities perceived by individuals with these conditions are simulated. As can be seen by comparing the colors perceived by individuals with these conditions with the colors perceived by individuals with normal trichromatic vision (top of Fig. 2.1), dichromats perceive fewer distinguishable color qualities. Some wavelengths that are associated with yellow or cyan in trichromats are indistinguishable from white for dichromats. Furthermore, dichromats cannot distinguish some wavelength pairs that are easily distinguished by trichromats. In protanopia, some colors that are perceived as green, yellow, or red by those with normal color vision are indistinguishable. Individuals with tritanopia cannot discriminate what in the normal case is perceived as yellow from what in the normal case is perceived as pink. If one would construct the color spaces for dichromats based on similarity relations between perceptual qualities, these spaces would be different from the normal color space shown in Fig. 1.1. However, like colors

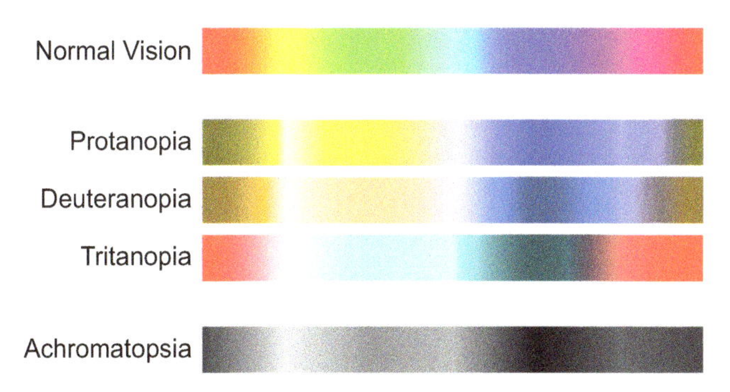

Fig. 2.1 Variability in color perception between individuals. How lights along the frequency spectrum are perceived by humans with different systems for color perception is shown. A color blindness simulator was used to generate this figure

perceived by normal perceivers, the colors that are perceived by dichromats all have hue, saturation, and brightness. Each perceptual color quality, regardless of whether it is perceived by human trichromats or dichromats, can be individuated by its position in the color space that has three dimensions representing hue, brightness, and saturation.

Dimensionality of Perceptual Spaces

Very rarely, individuals have even more impoverished color vision than dichromats. Individuals that suffer from total colorblindness, or achromatopsia, see the world only in different shades of gray. This condition can be caused by mutations that disrupt the molecular mechanisms with which cone cells respond to light. When this mechanism is disrupted, none of the three types of cone cells functions. Individuals with such mutations are not blind because in addition to cone cells, humans also have photosensitive rod cells. However, in the absence of functioning cone cells, humans only perceive different shades of gray. The visual perceptual qualities perceived by achromatopsic individuals all have the same hue and saturation. They differ from one another only in their brightness. The visual perceptual qualities of achromatopsic individuals can therefore be arranged in a one-dimensional space. The same is true for the visual perceptual qualities of monochromatic animals such as seals and owl monkeys.

Brightness is the single dimension of an achromatopsic individual's color space and also one of the three dimensions of the three-dimensional color space of individuals with normal vision. The one-dimensional achromatopsia color space is a subspace of the three-dimensional normal color space. The one-dimensional achromatopsia color space can therefore be transformed into a three-dimensional normal color space by adding two dimensions (saturation and hue) and assigning each perceptual quality the same value for these dimensions. A low-dimensional perceptual space that is a subspace of a higher-dimensional perceptual space can be represented in the same coordinate system as the higher-dimensional perceptual space. The difference in the coordinate systems is then reduced to a difference in the number and arrangement of perceptual qualities.

Since there are some species with lower dimensional color spaces than humans, it is likely that some species have visual perceptual quality spaces that have more than three dimensions. We know that many animals can perceive features of light that we cannot perceive. Some species, for example, can perceive the pattern of polarized light in the sky. When this pattern is visualized in textbooks, the differences of polarization angle are represented as differences in hue or brightness of the color of the clear blue sky, which looks homogeneous and uniform to us. However, we would have to construct the perceptual color space for such a species to find out how the polarization pattern is actually perceived by them. It is possible that polarization angles of light change the perceived hue, saturation, or brightness in species that are sensitive to polarization of light. However, it is also possible that the physical feature "angle of polarization" corresponds to a fourth dimension of color quality space that is inaccessible to humans. Similarly, we do not know whether pigeons, which are believed to be able to discriminate many more colors than humans, discriminate all colors depending on their hue, brightness, and saturation. Alternatively, the pigeon color space could have a fourth dimension that makes it easier to accommodate the large number of colors.

It would be arrogant to assume that no other species has a higher-dimensional color space than us. On the other hand, it is also a natural assumption. It is very easy to accept that the achromatopsic, who cannot perceive saturation or hue, is missing an important part of what makes visual perceptual qualities. However, it is much more difficult to imagine that we are also missing an important dimension of visual perceptual qualities. What else could there be to colors other than hue, saturation, and brightness? That this question is impossible to answer for somebody whose first-hand visual experience consists only of hue, saturation, and brightness illustrates the importance of relying on behavioral experiments rather than subjective experience in individuating perceptual qualities.

2.2 A Strategy to Compare Perceptual Qualities Perceived by Different Perceivers

Different perceivers often perceive similarities between perceptual qualities differently and the perceptual spaces that reflect these differences are therefore different. Because of this variability in perception the question whether another individual perceives the color and smell of a ripe tomato in the same way I perceive them has no general theoretical answer. However, comparing perceptual qualities between perceivers is necessary for third-person access to perceptual qualities, which, in turn, is necessary for an objective science of perceptual qualities.

Although perceptual qualities that are perceived by different perceivers cannot be compared based on theoretical considerations, an empirical approach makes such comparisons possible in many cases. The question how similar two perceivers' perceptions of a ripe tomato are can therefore be answered objectively.

I will show how this question can be answered in some cases by constructing and then registering the perceptual spaces of the two perceivers under study. This is possible in many situations that are of great theoretical interest, for example, when comparing perception in two healthy individuals of the same species. However, I will also show that comparing perceptual qualities between perceivers is only possible when the two perceivers' perceptual spaces can be transformed, either directly or through intermediates, into the same coordinate system.

Perceptual Space Registration

How can we measure the similarity between perceptual qualities that are perceived by different perceivers? The similarity between perceptual qualities that are perceived by the same perceiver is represented by the distance between the two perceptual qualities in the exhaustive perceptual space. However, as discussed above, perceptual spaces differ between perceivers. This introduces a problem for comparing perceptual qualities that are not perceived by the same perceiver. As an illustration, consider a city map. I can mark locations on a city map and then compare one location to another. When the marks are close on the map, the positions they represent are geographically close. When the marks are far apart, then they represent geographically distant locations. Two marks at the same spot on the map represent the same location in the city. Now imagine that a friend of mine has the same city map and that she marks locations on her map. It will be easy, at least theoretically, to compare locations marked on the two maps. All that needs to be done is to transfer the marks from one map to the other identical map and then measure the distances between them. This is analogous to comparing perceptual qualities between two perceivers that have the same perceptual space. However, as reviewed above, no two perceivers have the same perceptual space. The analogous situation is therefore a situation in which different locations have been marked on two different maps of the city. Imagine that one map is a 1:10,000-scale map and the other a 1:20,000-scale map. In this situation, the two maps have to be registered before locations that are marked on the different maps can be compared. In the case of a 1:10,000-scale map and 1:20,000-scale map, the registration algorithm simply consists in rescaling one of the maps. The solution to the problem of comparing perceptual qualities in different perceptual spaces is similar to the problem of comparing locations marked on different maps: the data points that are to be compared have to be transformed into the same coordinate system before they can be compared.

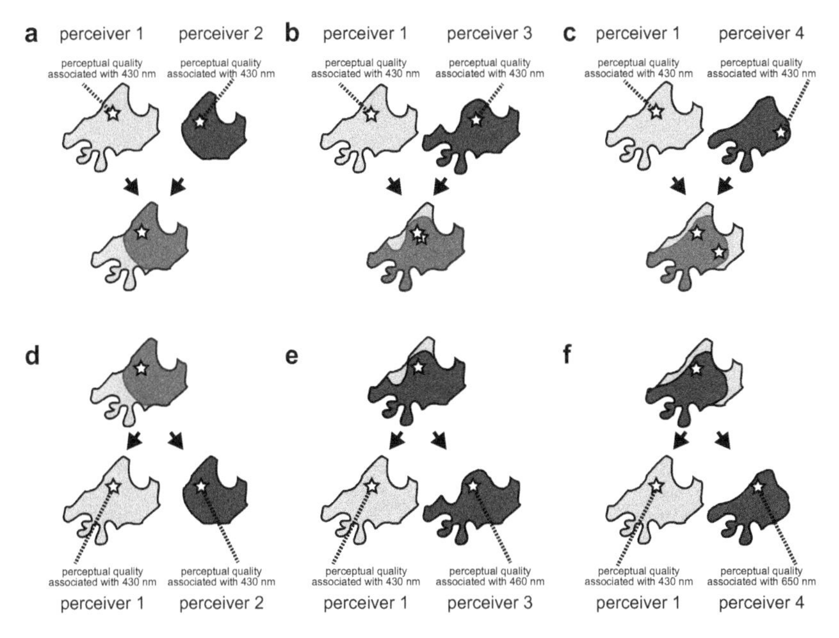

Fig. 2.2 Determining the similarity of perceptual qualities perceived by different perceivers. (**a–c**) The perceptual quality associated with light of 430 nm wavelength in perceiver 1 is compared to the perceptual quality associated with light of 430 nm wavelength in three other perceivers. (**a**) In perceivers 1 and 2, the same perceptual quality is associated with light of 430 nm wavelength. (**b**) In perceivers 1 and 3, a similar perceptual quality is associated with light of 430 nm wavelength. (**c**) In perceivers 1 and 4, very different perceptual qualities are associated with the same stimulus. (**d–f**) The stimulus that is associated with a specific perceptual quality in perceiver 1 is compared to the stimuli associated with the same perceptual quality in three other perceivers. (**d**) In perceivers 1 and 2, the same stimulus (light of 430 nm wavelength) is associated with the perceptual quality. (**e**) The perceptual quality associated with light of 430 nm wavelength in perceiver 1 is associated with light of 460 nm wavelength in perceiver 3. (**f**) The perceptual quality associated with light of 430 nm wavelength in perceiver 1 is associated with light of 650 nm wavelength in perceiver 4

Consider a hypothetical example. In the top row of Fig. 2.2a–c, the two perceptual spaces of two different perceivers and the position of the perceptual quality associated with a light stimulus of 430 nm in each of the spaces are shown. The question we want to answer is whether the perceptual quality that is associated with a light stimulus of 430 nm is the same for both perceivers. To answer this question, the two perceptual spaces have to be registered based on their structure (lower row in Fig. 2.2a–c).[2] When the two spaces are registered, the distance between the perceptual qualities associated with light of 430 nm wavelength in the two perceivers can be measured. In Fig. 2.2a, the perceptual qualities of perceiver 1 and 2 are at the same position in the combined perceptual space, which means that the same perceptual quality is associated with this stimulus in those two perceivers. In Fig. 2.2b, c, different perceptual qualities are associated with the same stimulus in different perceivers. The perceptual qualities associated with light of 430 nm wavelength in perceiver 1 and 3 are similar (Fig. 2.2b), whereas radically different perceptual qualities are associated with this stimulus in perceivers 1 and 4 (Fig. 2.2c).

Another question one can ask is which stimuli elicit the same perceptual qualities in two perceivers. The strategy to answer this question is illustrated in Fig. 2.2d–f. To answer this question, the two perceptual spaces have to be registered and the position of the perceptual quality under study in them has to be marked (upper row in Fig. 2.2d–f). Then, the physical stimulus that elicits this perceptual quality can be identified (lower row in Fig. 2.2d–f). For perceivers 1 and 2, the same stimulus elicits the perceptual quality under study (Fig. 2.2d). In perceiver 3, light of 460 nm wavelength elicits the same perceptual quality that is elicited by light of 430 nm wavelength in perceiver 1 (Fig. 2.2e). In perceiver 4, light of 650 nm wavelength elicits the same perceptual quality (Fig. 2.2f).

It is important to understand that the physical nature of the stimulus cannot play a role in registering perceptual spaces. It may be tempting to use physical stimuli that are perceived by both perceivers to anchor the two perceivers' perceptual spaces. One could, for example, attempt to register the perceptual color space of a bee and the perceptual color space of a human by distorting them so that the perceptual qualities that are associated with 480 nm light (blue for humans), 560 nm light (green for humans), and 710 nm light (red for humans) are in the same position. Then, one could identify the position of the perceptual quality associated with UV light (which can be perceived by bees, but not by humans) in the bee's perceptual space and measure what the closest perceptual quality in the human's perceptual space is. This, one might believe, would be a way to identify the human color perception that is most similar to how bees perceive UV light. However, this strategy is flawed because it assumes that the perceptual qualities blue, green, and red are elicited by the same stimuli in humans and bees, and there is no evidence supporting this assumption.

Let us return to the city map analogy. That the physical nature of the stimulus cannot play a role in registering perceptual spaces is analogous to saying that the marks on the city maps cannot play a role in registering the maps. Imagine two different maps of the city. On one, I have marked the location of my home, my workplace, my barber, and my favorite restaurant. On the other, you have marked your home, your workplace, your barber, and your favorite restaurant. We want to know how close our favorite restaurants are. We cannot achieve this by transforming one of the maps so that the marks marking your home and my home, your workplace and my workplace, and your barber and my barber are at the same position and then measuring the distance between the marks marking your favorite restaurant and my favorite restaurant. The marks on the maps cannot be used to register the maps. Only features of the maps themselves can be used to register them. We have to transform one of the maps so that the railway station, the airport, and the opera house are in the same position in both maps. Then, the distance between the marks marking your and my favorite restaurant can be measured. In analogy, the only way to find out which of your perceptual qualities is most similar to the perceptual quality associated with UV light in bees is to construct your own perceptual space and the perceptual space of the bee. If the two spaces can be registered based on their structure, then the position of the perceptual quality associated with UV light in the bee can be compared to your perceptual space and the perceptual quality in your perceptual space closest to it can be identified.

Examples of Third-Person Access to Perceptual Qualities

Consider, as an example of how two perceptual spaces can be registered, the perceptual space of someone who suffers from specific anosmia for musks. Specific anosmias are conditions in which people with an otherwise normal sense of smell cannot smell a certain type of odor, for example, musk odors (Amoore 1967). This condition can be caused by a disruptive mutation in the odorant receptor sensitive for these odors (Whissell-Buechy and Amoore 1973; Keller et al. 2007; Mainland et al. 2014). How could somebody with specific anosmia to musk find out something about the perceptual qualities that are associated with musk? Knowledge about the musk-sensitive odorant receptors and about the musk molecules is not going to be helpful because neither similarity between odorous molecules nor that between odorant receptors is a reliable predictor of similarities between olfactory perceptual qualities. The way to learn about the perceptual qualities of musk is therefore to build two smell spaces, one for an individual with a normal sense of smell and one for an individual with specific musk anosmia. The two smell spaces will be almost identical except that one will have a hole at the position at which the musk perceptual qualities are located in the smell space of the normal subject (Fig. 2.3a, left). This hole represents the blind spot for musks and it is surrounded by the perceptual qualities that constitute the border of the hole. The smell of musk is between the smell of perceptual qualities on opposing sides of the hole that represents the musk blind spot. The individual with musk anosmia can find out what the physical stimuli are that are associated with the perceptual qualities around the blind spot and then smell them.

The advantage of specific anosmia as an example is that it is an actual condition. The disadvantage is that the smell space is not known yet and the example can therefore only be discussed in abstract terms. To complement the example of specific anosmia, imagine a fictional example that is similar to Hume's *Missing Shade of Blue*. Imagine someone having a

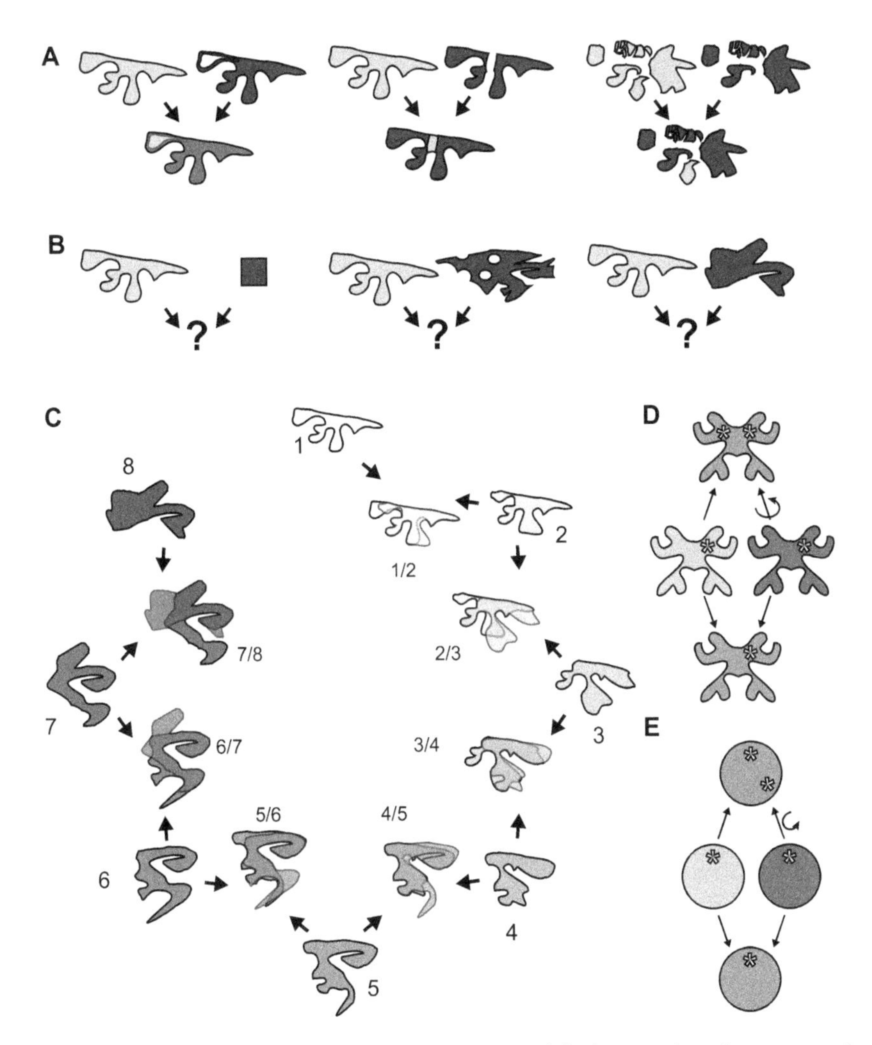

Fig. 2.3 Registering perceptual quality spaces. (**a**) Three pairs of perceptual spaces that are sufficiently similar in their shape to be registered. (**b**) Three pairs of perceptual spaces that are too dissimilar to be unambiguously registered. (**c**) Perceptual space 1 and perceptual space 8 are not similar enough to be directly registered. However, through a chain of intermediates (perceptual spaces 2 to 7) they can be registered indirectly. (**d, e**) Symmetrical perceptual spaces can be registered in more than one orientation

normal auditory system with the exception of the inability of hearing the pitch D3. When somebody is playing a scale on a piano, there will be a moment of silence between C3 and E3 for this person. People with an intact auditory system will hear D3 between C3 and E3. The tone space is two dimensional. One dimension is pitch and the other loudness. D3-deaf people will not hear anything at the pitch D3 at any loudness, so the tone space of the D3-deaf person will differ from that of a person with normal hearing by lacking a stripe along the pitch dimension. The tone space will be cut into two tone spaces, one space for tones higher than D3 and one space for tones lower than D3 (Fig. 2.3a middle). Because in a scale, the perceptual qualities are ordered along the dimension (pitch) in which there is the specific blindness, the note before and after the moment of silence will provide information about the perceptual quality of D3 to the D3-deaf subject.

Consider a third example. People suffering from ageusia have no sense of taste. Constructing a perceptual quality space of such an individual will result in a space in which one modality-representing cluster, the one that represents taste qualities, is missing. It will likely be possible to register such a perceptual quality space with the perceptual quality space of a person that has all senses (Fig. 2.3a right). Let us assume that there is no continuum of perceptual qualities between the gustatory qualities and qualities in other modalities, so there are no perceptual qualities that are directly adjacent to the missing taste qualities. However, there are proximity relations between perceptual qualities in non-gustatory modalities in the ageusia patient's perceptual space and taste qualities in the normal perceptual space that can be registered with it. Unless the experiments are performed and the exhaustive perceptual spaces are constructed, the nature of the proximity relations between these perceptual qualities is unknown. However, it seems likely that the registration will reveal something about gustatory perceptual qualities to the ageusia patient. I would speculate that the gustatory perceptual quality associated with sugar is closer to the perceptual qualities associated with vanilla smell and soft touch perceived at the tongue than it is to the perceptual quality associated with the color red.

These examples show that the strategy for third-person access to perceptual qualities outlined above can be successful. It is possible for the individual with musk anosmia to have information about how an individual with a normal sense of smell experiences musks. There is no ontological or epistemological barrier that precludes us from knowing other minds. This is a very important result because without a way to compare perceptual qualities regardless of who perceives them, there could be no objective science of perception. The fact that nothing precludes knowledge of other minds in principle does not mean that there are no practical problems with third-person access to perceptual qualities. In the next section, I will discuss these practical problems and show that some of them are insurmountable and therefore pose limits for third-person access to perceptual qualities.

2.3 Limits of Third-Person Access to Perceptual Qualities

The strategy for third-person access to perceptual qualities outlined above depends on registering the perceptual spaces of the two perceivers whose perceptual qualities we want to compare. Registering perceptual spaces, however, may not always be possible. In other situations, there may be more than one possible registration, so the two perceptual spaces cannot be registered unambiguously.

Perceptual Spaces That Cannot Be Registered

In general, the more similar two spaces are, the easier it is to register them. In Fig. 2.3a, three hypothetical examples of pairs of perceptual spaces with sufficient similarity in shape to register them are shown. The perceptual space pairs correspond to the examples of musk anosmia (left), D3-deafness (middle), and ageusia (right). In each of these three examples, the space on the right is derived from the space on the left by subtracting a small area. This makes it easy to register the two spaces. In Fig. 2.3b, three pairs of perceptual spaces that are so dissimilar that it is not possible to register the light gray space with the dark gray space unambiguously are shown. In most cases in which two perceptual spaces have to be registered, the decision whether it is possible to register them unambiguously or not is presumably not as clear cut as in the examples in Fig. 2.3a, b. The perceptual spaces of two human individuals with normal perceptual capacities will presumably have very similar overall shapes. However, there will also be differences between the spaces. Maybe the cluster representing colors is slightly elongated in one of the spaces, and the distance between the cluster of smells and tastes is not the same. It will certainly be possible to register the two exhaustive perceptual spaces, but it may be that several, slightly different registrations are possible. One then has to use an algorithm to decide which registration is the best. Another complication for deciding whether two spaces can be registered is that it has to be decided which transformations of the spaces should be allowed to register them. If any type of transformation is acceptable, then every space can be distorted so that it will perfectly register with every other space. Which algorithm is chosen for registering the perceptual spaces, and which types of transformations are permitted are arbitrary decisions. There can therefore be cases in which it is disputable whether two exhaustive perceptual spaces can be registered or not. However, there will also be cases in which it is clear whether two perceptual spaces can be registered or not.

When two perceptual spaces that clearly cannot be registered directly are found in different animal species, a strategy to register them despite their very different structures is sometimes available. The strategy is to find intermediates that can be used to create a chain of registered perceptual spaces. The two dissimilar perceptual spaces 1 and 8 in Fig. 2.3c cannot be registered directly. They are the same pair of perceptual spaces shown on the right in Fig. 2.3b as an example of spaces that cannot be registered. However, as is shown in Fig. 2.3c, 1 and 8 can be registered indirectly through the series of perceptual spaces 2–7. Perceptual space 1 can be registered with perceptual space 2, perceptual space 2 can be registered with perceptual space 3, and so on, until spaces 1–8 form a chain of unambiguously registered perceptual spaces. The direct registration of 1 and 8 has not been possible because of the different shapes of the two perceptual shapes. The indirect registration of 1 and 8 through a series of intermediates has, however, been possible. To provide a hypothetical example, it may not be possible to register the human perceptual space directly with the perceptual space of the mouse. However, if the human perceptual space can be directly registered with the monkey perceptual space, and the monkey perceptual space can be directly registered with the rabbit perceptual space, and the rabbit perceptual space can be directly registered with the mouse perceptual space, it would still be possible to compare the perceptual qualities associated with 480 nm light in mice and in humans.

Perceptual Spaces That Can Be Registered in Different Orientations

Perceptual spaces that are too different to be registered are not the only problem for comparing perceptual qualities between different perceivers. Another potential problem is that symmetrical perceptual spaces can be registered in more than one orientation (Fig. 2.3d, e). Mirror symmetric perceptual spaces can be registered in two different orientations (Fig. 2.3d). Radial symmetric perceptual spaces can be registered in any orientation (Fig. 2.3e). Symmetry is more likely in simple perceptual spaces. In humans, neither the three-dimensional color space nor the two-dimensional tone space is symmetrical. The structure of perceptual spaces in other modalities is not well known. Maybe there is a symmetrical one-dimensional temperature space that goes from hot to cold. The exhaustive perceptual space that contains all perceptual qualities is complex and contains the non-symmetrical color and tone spaces, so the worry about symmetry is not pressing for the account of human perceptual qualities that I propose. However, if a symmetrical perceptual space were to be discovered in an animal species, the correct orientation could probably be found by studying the perceptual space of closely related species with a similar, yet not symmetrical, perceptual space. The most likely situation in which symmetry of the space will limit third-person access to perceptual qualities is when it comes to the perceptual spaces of machines, which can be purposefully designed to have symmetrical perceptual spaces.

In the philosophical literature, discussions of the possibility of symmetrical perceptual quality spaces has a long tradition since John Locke's discussion of the possibility of spectrum inversion in color perception (Locke 1689/1975) (for an overview of the literature, see Byrne 2010). Locke, writing long before the human perceptual color space had been established, discussed the hypothetical case of a symmetrical color space. In the case discussed by Locke, strawberries elicit the perceptual qualities that are elicited by cucumbers in normal perceivers. Cucumbers elicit the perceptual qualities normally elicited by strawberries. Locke goes on to assure the reader that despite this possibility of an inverted color spectrum, he thinks that the same object will elicit the same, or very similar, mental qualities in different individuals. Because the perceptual spaces of humans are not symmetrical, it is now possible to empirically test this proposal and search for individuals who perceive as green what others perceive as red. John Dalton discovered more than hundred years after Locke's hypothetical case that some individuals are red-green colorblind. In these individuals, cucumbers and strawberries elicit some of the same perceptual qualities.

Because the exhaustive perceptual quality space containing all perceptual qualities is not symmetrical, undetectable inversion is not possible in humans. However, it is possible in creatures with symmetrical perceptual quality spaces (if such creatures exist). This theoretical possibility of undetectable spectrum inversion is sometimes used in arguments against behaviorism and functionalism (Fodor and Lepore 1996). The argument is that, in undetectable spectrum inversion, two behaviorally or functionally identical perceivers could perceive different perceptual qualities in identical situations. It would be impossible to prove that there are no cases of spectrum inversion (hence "undetectable" spectrum inversion).

Providing convincing evidence for the absence of something is notoriously difficult. However, this is not a peculiarity of perceptual research, but applies widely. That there are limits to the third-person access to perceptual qualities should not be mistaken for a profound discovery that demonstrates that perceptual qualities are different from other things considered by science. Limits to what can be known are common in science. Consider, for example, that part of the universe is causally disconnected from us because it is so far away that light did not have enough time to travel from there to the earth since the big bang. If this part of the universe is furthermore continually moving away from us with the speed of light, then it is beyond the "future visibility limit", and it will forever stay causally disconnected from us. We can never know anything about this part of the universe. However, it does not follow from the fact that there is some matter that cannot be observed that it is not possible to study matter scientifically. Similarly, it would not follow from the fact that a perceiver with a symmetrical perceptual space exists that functionalism is false.

2.4 Conclusion: Registering Perceptual Spaces Enables Third-Person Access to Perceptual Qualities

Different perceivers perceive stimuli and the similarity between stimuli with different perceptual systems and they therefore have different perceptual quality spaces. This diversity of perceptual spaces complicates comparisons between perceptual qualities that are perceived by different perceivers. I propose a strategy for third-person access to perceptual qualities that is based on registering perceptual spaces. Quantifying the distance between perceptual qualities that are perceived by two different perceivers is possible by registering the two perceptual spaces and then measuring the distance between the perceptual qualities in the combined space. This objective measure of similarity between perceptual qualities regardless of whom they are perceived by provides the basis for the scientific study of perceptual qualities.

Because the strategy for third-person access depends on the ability to unambiguously register the perceptual spaces of the two perceivers that are compared, it is not always possible. One problem is that some perceptual spaces have structures that are so different that they cannot be registered. The structure of the perceptual space of a perceiver depends on its perceptual systems and on the processing of sensory information. The perceptual systems of two biological perceivers are connected through the perceptual system of their last common ancestor. It is therefore unlikely to find perceptual spaces so radically different that they cannot be registered among close relatives. It is also possible, when two perceptual spaces cannot be registered directly, to use a chain of intermediates to register them indirectly. However, this is not always possible. Humans are evolutionarily so distant from bacteria that it will very likely not be possible to register the human perceptual space through intermediate steps with the perceptual space of a bacterium. It is probably not possible to compare a given human perceptual quality with a given bacterial perceptual quality. The situation becomes even more hopeless when the two perceivers that are to be compared are not connected through a common ancestor. The perceptual space of an animal is unlikely to be similar enough to the perceptual space of a robot or of an extraterrestrial for the two spaces to be registered unambiguously. It will therefore not be possible to find a stimulus that is associated with the same perceptual quality in a robot and in a bat.

Another problem case for the strategy of comparing perceptual qualities between perceivers by registering the perceivers' perceptual spaces is symmetrical perceptual spaces. Symmetrical spaces can be registered in more than one orientation. The exhaustive perceptual space in humans is not symmetrical, so this is not a concern when it comes to comparing our own perceptual qualities to those of other human perceivers. However, it is possible that some perceivers have symmetrical perceptual spaces.[3] It is most likely that symmetrical perceptual spaces will be found in very simple organisms, or in machines, which can be purposefully designed to have symmetrical perceptual spaces.

Notes

1. Having more types of receptors with different sensitivities does not always lead to more discriminable perceptual color qualities. The direct relation between the number of different receptors and the number of discriminable perceptual qualities only holds when the information collected by the receptors is processed in the same way. This has been illustrated by research into species of mantis shrimp that have up to 12 different photoreceptors, each sampling a different narrow range of wavelengths. Marshall, J., T. W. Cronin, et al. (2007). "Stomatopod eye structure and function: A review." *Arthropod Structure & Development 36*(4): 420–448. Based on the large number of photoreceptors with different sensitivities, it has been speculated that mantis shrimp have a very large and complex color space. However, behavioral experiments revealed that this is not the case, which led to the suggestion that mantis shrimp do not use comparisons between different channels to discriminate between colors. Thoen, H. H., M. J. How, et al. (2014). "A Different Form of Color Vision in Mantis Shrimp." *Science 343*(6169): 411–413. Assuming that the processing of color

information is relatively conserved, a relation between the number of opsin genes and the number of colors that the species can discriminate can be expected in vertebrates.

2. The two spaces have to be aligned depending on their structure not depending on their dimensions. As Jerry Fodor and Ernie Lepore discussed in the context of State Space Semantics Fodor, J. and E. Lepore (1996). Paul Churchland and State Space Semantics. *The Churchlands and their critics*. R. N. McCauley. Cambridge, Blackwell Publishing: 145–159., attempts to align similarity spaces based on their dimensions merely succeed in shifting the problem from third-person access to perceptual qualities to third-person access to the dimensions of perceptual spaces. Instead of wondering whether your "blue" is the same as my "blue", we would wonder whether your "saturation" is the same as my "saturation". Aligning the spaces depending on their structures avoids this problem.
3. Rosenthal argues that it is impossible for a space that is based on an individual's discrimination abilities to be symmetrical because a symmetrical space would collapse around the axis of symmetry. Rosenthal, D. M. (2010). "How to think about mental qualities." *Philosophical Issues: Philosophy of Mind 20*: 368–393. This is not the case for the perceptual spaces discussed here. To illustrate that symmetrical perceptual spaces are possible, let us consider a simple perceiver that can only perceive two discriminable perceptual qualities. The perceptual space of this perceiver would consist of two points that represent the two perceptual qualities. This space would be symmetric because it is always possible to draw a symmetry axis between two points. If one would collapse this space around the axis of symmetry, the two perceptual qualities would fall into the same location. Since the two perceptual qualities can be discriminated, the collapsed space (which represents the two qualities as identical) would not represent the similarity relations between the perceptual qualities accurately.

References

Amoore, J. E. (1967). Specific anosmia: A clue to the olfactory code. *Nature, 214*(5093), 1095–1098.

Byrne, A. (2010). Inverted Qualia. *The Stanford Encyclopedia of Philosophy* Spring 2010. Retrieved June 23, 2011, from http://plato.stanford.edu/archives/spr2010/entries/qualia-inverted

Fodor, J., & Lepore, E. (1996). Paul Churchland and state space semantics. In R. N. McCauley (Ed.), *The Churchlands and their critics* (pp. 145–159). Cambridge: Blackwell Publishing.

Jacobs, G. H. (1981). *Comparative color vision*. New York: Academic.
Jacobs, G. H. (2009). Evolution of colour vision in mammals. *Philosophical Transactions of the Royal Society, B: Biological Sciences, 364*, 2957–2967.
Keller, A., Zhuang, H., et al. (2007). Genetic variation in a human odorant receptor alters odour perception. *Nature, 449*(7161), 468–472.
Locke, J. (1689/1975). *Essay concerning human understanding*. Oxford: Oxford University Press.
Mainland, J. D., Keller, A., et al. (2014). The missense of smell: Functional variability in the human odorant receptor repertoire. *Nature Neuroscience, 17*(1), 114–120.
Marshall, J., Cronin, T. W., et al. (2007). Stomatopod eye structure and function: A review. *Arthropod Structure & Development, 36*(4), 420–448.
Thoen, H. H., How, M. J., et al. (2014). A different form of color vision in Mantis Shrimp. *Science, 343*(6169), 411–413.
Whissell-Buechy, D., & Amoore, J. E. (1973). Odour-blindness to musk: Simple recessive inheritance. *Nature, 242*(5395), 271–273.

Part 2

Percepts

In the first part of this book, I have discussed perceptual qualities abstracted away from the spatial and temporal structure of perception. Perceptual qualities are the basic building blocks of perception. However, normal instances of perception are much richer and more complex than a series of changing perceptual qualities. In this second part, I will discuss percepts, which emerge through the combination of perceptual qualities. In Chap. 3, the notion of objecthood in the context of perception will be examined, and in Chap. 4, I will discuss the function of perception. Specifically, I will argue against the proposal that it is the function of perception to collect correct information about the physical world. Instead, I will show that the function of perception is to guide behaviors.

In this part, I will cash in on the promise from the book's title and focus on olfactory perception. I hope that it will become apparent that notions like object-oriented perception and the idea of perception as a mirror of reality, which seem intuitive when considering visual perception, fail to capture the nature of olfactory perception.

3

Olfactory Objects

It usually does not appear to us as if we perceive perceptual qualities. Instead, we seem to perceive objects and their properties. Heidegger wrote, "Much closer to us than any sensations are the things themselves. In the house, we hear the door slam – never acoustic sensations or mere noises" (Heidegger 2002/1950, p. 8). Based on this introspective evidence, object-based accounts of perception have been developed. Objects of perception are most often associated with vision and touch. The traditional auditory counterpart to the visual object is the auditory event (Blauert 1997). In the name of identifying commonalities between modalities, talk of "auditory objects" has become more common both in neuroscience (Griffiths and Warren 2004; Bizley and Cohen 2013) and in the philosophy of perception (O'Callaghan 2008; Matthen 2010; Nudds 2010).[1]

The modality to which an object-based account of perception has been most difficult to generalize is olfaction.[2] Barwich writes: "The more we understand about the multidimensionality of olfactory experiences and the processes of smell perception with which they resonate, the more apparent it should become that object-based talk about perceptions is no longer tenable, if it ever was" (Barwich 2014). In this chapter, I will

A. Keller, *Philosophy of Olfactory Perception*,
DOI 10.1007/978-3-319-33645-9_3

explain why I agree with Barwich. Because there is a close connection between notions of objecthood and spatial location and extension, I will first review the role of space in olfactory perception. Based on this review, I will then show that the common criteria for objecthood in vision fail to pick out olfactory objects. Finally, I will discuss the proposed candidate olfactory objects and argue that neither of them justifies applying the notion of perceptual objects to olfaction.

3.1 Olfaction in Space

Introspectively, it does not seem that the perception of odors is spatially structured. Olfactory perception has been called "a single unitary experience" (Stevenson 2009, p. 1010), "a state of consciousness, having neither geometry nor articulate individuation" (Lycan 2000, p. 282), and a "nominal sense" that simply provides "data of qualitatively different odors in our surroundings" (Köster 2002, p. 28). On the other hand, humans can indisputably use olfactory information for navigation. This is only superficially a contradiction because, as I will explain below, the ability to navigate using olfactory stimuli does not imply that olfactory perception is spatially structured.

Spatial Structure of the Olfactory Environment

Our olfactory environment is spatially structured at different scales. At the largest scale, New Delhi, the Bavarian Forest, the Mexican resort town of Cancún, and the Mojave Desert in the southwestern USA can all easily be distinguished from one another based on the odorous molecules in the air at these locations. Each of these places has a different distinct smell. On a smaller scale, different ecosystems also have different smells. The ocean smells different from the tidal zone, which smells different from the beach, which smells different from the marshland. If one walks toward the ocean from a few miles inland, the odor environment will change in a predictable pattern. On a yet smaller scale, the bakery in the mall smells different from the shoe store. At an even smaller scale, my apartment smells different from the hallway and my neighbors' apartments. Within my apartment, the kitchen, the bathroom, and the bedroom all have unique smells. Within the kitchen, the area by the oven smells different from the sink or the trashcan.

The spatial distribution of odorants in the world is due to the spatial arrangement of the odor sources that release characteristic odorants. The concentration of the released molecules is, on average, higher closer to the source. Some odor sources remain at the same position in the environment and give off the same odorants for a long time. The ocean, for example, is a reliable olfactory landmark. Many other odor sources are much less permanent and reliable. The bakery in the mall will stop giving off bakery smells when it closes for the night. Apple trees may be good olfactory landmarks when they are flowering and then again when the apples are rotting under them, but not during the winter months. The difference between olfactory landmarks and visual landmarks, like rivers, buildings, and mountains, is that olfactory landmarks are less stable over time.

The distribution of odorous molecules in the environment is especially unstable at the scale between meters and kilometers that is behaviorally relevant to humans. At this scale, the spatial distribution of odorous molecules is determined mostly by turbulent airflow (Weissburg 2000). At a lower scale between millimeters and centimeters, odor distribution is mainly determined by diffusion. The diffusion of odorants results in the formation of gradients and odor gradients can be used to locate the odor source. This is how insect larvae use smell for navigation. At a larger scale of tens or hundreds of kilometers, odor distribution is determined by climate events that produce stable spatial gradients in ratios of atmospheric trace gases (Wallraff and Andreae 2000). These stable gradients at a kilometer scale allow homing pigeons to smell their way home (Gagliardo 2013). At the human scale, stable patterns of odor distribution are comparatively rare, even in the presence of stable odor sources.

Dogs, rats, and other animals successfully use odors for navigation, which shows that, at least sometimes, odorants have stable spatial distributions at the scale that is behaviorally relevant for humans. However, dogs, rats, and other terrestrial mammals normally use odors deposited on the ground as trails or markings. Trails and markings are much more stable over time than clouds of molecules. However, trails and markings are not accessible to humans because with the development of bipedalism our noses have been moved too high above the ground. Other species found solutions for similar problems. The elephant's olfactory epithelium is far from the ground, but it compensates for this with its trunk that allows it to suck the odorous molecules from the ground up to the nasal cavity. Humans may be the only species of terrestrial mammals that is not aware of the trails and markings that structure the ground they walk on.

In summary, in the physical world, some non-random distributions of odorous molecules are stable over certain periods, for example, odor gradients at the centimeter scale in soil. However, in the air several feet above ground, where our noses are, the distribution of odorous molecules is generally strongly influenced by turbulent airflows and very unreliable.

Spatial Structure of Olfactory Perception

Our odor environment has a spatial structure and it would be possible that this structure is reflected in the structure of olfactory perception in the way in which the spatial structure of a landscape is reflected in visual perception. However, laboratory experiments have shown that human olfactory perception is spatially unstructured. The simplest possible spatial structure of olfactory perception, a division of the perception into left nostril and right nostril, would be revealed by the ability to tell whether an odor is applied to the left or to the right nostril. Subjects fail even at this simplest task. It is usually not possible to discriminate between a stimulus being presented in the left nostril and the same stimulus being presented in the right nostril (Radil and Wysocki 1998; Frasnelli et al. 2008).

In seeming contradiction, it has also been reported that humans can compare the olfactory input between their two nostrils (von Békésy 1964; Porter et al. 2005, 2007). The most likely explanation for this discrepancy is that the performance in experiments in which the correct nostril was identified is not based on olfaction, but on chemesthesis (Kleemann et al. 2009; Croy et al. 2014). Chemesthesis is a different modality than olfaction.[3] The mucous membrane inside our nasal cavity has chemesthetic sensitivity. Like the entire surface or our body, this membrane is spatially mapped in our brain and we can tell whether the skin in our left or in our right nostril is being irritated. Experimentally dissociating olfaction and chemesthesis is notoriously difficult because at high concentrations many volatile molecules activate both the olfactory system and the trigeminal nerve. However, other than sharing the same stimuli, olfaction and chemesthesis have little in common (Fig. 3.1).

Odor-Guided Navigation Without Spatially Structured Perception

A possible objection to the thesis that our olfactory perception has no spatial structure is to point out that humans can successfully navigate using olfactory information. It is possible to find the source of an odor in a room. When humans navigate toward an object based on visual or auditory information, they use the spatial structure of the visual or auditory perception for guidance. The parsimonious assumption is that humans use the same strategy when using odor information to navigate toward an object. However, investigating instances of odor-guided navigation shows that this is not the case. Humans use different strategies for navigating based on olfactory perception than for navigating based on perception in modalities in which perceptions have spatial structure.

Although humans do not routinely use olfaction for navigation, there can be no doubt that it is possible. The paradigm example for olfactory navigation in humans is the nipple search behavior, in which infants use olfactory cues to orient toward their mothers' breasts (Varendi et al. 1994; Varendi and Porter 2001). There are also other examples. In dense tropical forests, visual navigation can be difficult, but there are many distinct odor sources available for olfactory navigation. Consequently, people living in this environment sometimes develop smell maps which they use for navigation (Classen et al. 1994, pp. 97–99). Smell can also provide valuable information about one's position with respect to an odor source, for example, a lion. One can, using olfactory information, avoid areas that have a strong lion smell. Undoubtedly, this ability has saved many human lives.

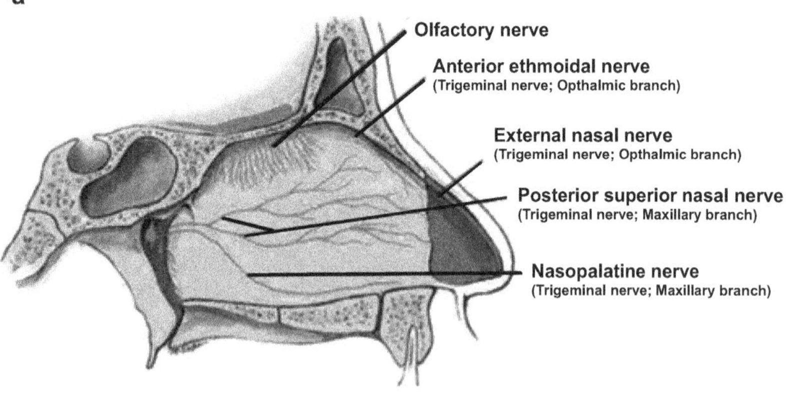

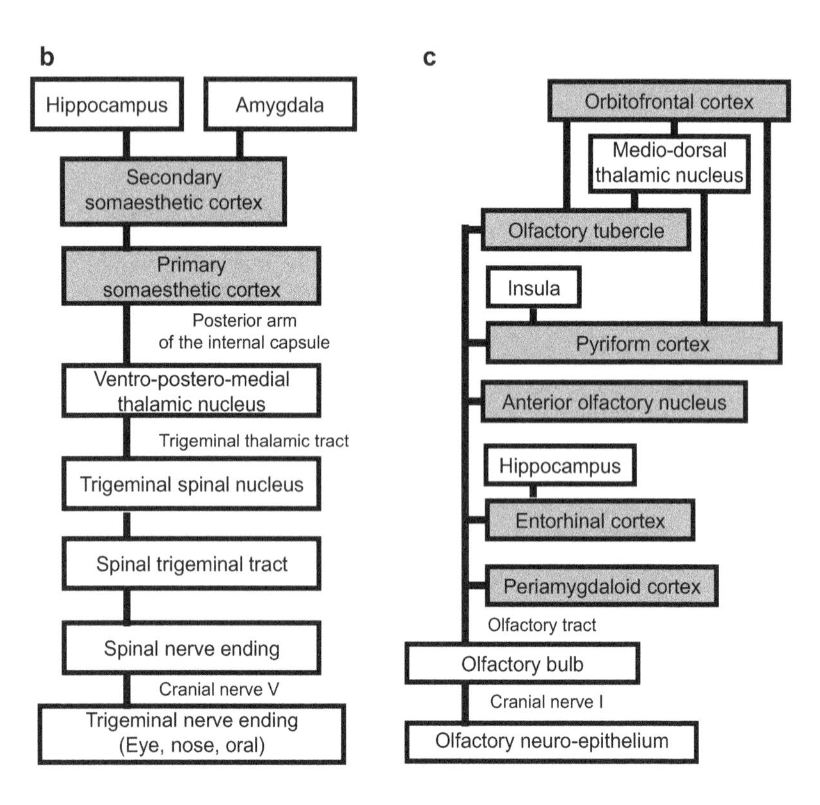

Fig. 3.1 Perception mediated by the olfactory and by the trigeminal nerve. (**a**) The olfactory nerve innervates the olfactory epithelium in the nasal cavity, whereas branches of the trigeminal nerve innervate the respiratory epithelium in the nasal cavity. (**b, c**) Different brain areas are involved in the processing of trigeminal information (**b**) and of olfactory information (**c**). (in (**b**) and (**c**) cortical regions are in *gray* and non-cortical regions in *white*)

These observations show that humans can use their sense of smell for navigation. That humans have the ability to navigate using odors has also been shown experimentally. Blindfolded human subjects have the ability to follow, crawling on all fours, a scent trail of chocolate essential oil (Porter et al. 2007) (Fig. 3.2a). In another experiment, subjects were shown to be able to determine their position with respect to two odor sources by analyzing the ratio of the two smells (Jacobs et al. 2015) (Fig. 3.2b). In institutions for mentally or visually handicapped people in the Netherlands, odorizing different hallways with different odors reduced the number of inhabitants that got lost trying to find their rooms (Köster et al. 2014).

That humans can use olfactory perception to navigate is not surprising. All perception of stimuli that are spatially structured in the environment provides the necessary information for navigation. If, in the experiment by Porter and colleagues (Porter et al. 2007), a salt trail would have been laid on the ground instead of a chocolate odor trail, blindfolded subjects presumably could have licked their way along the salt trail. In a gustatory version of the study by Jacobs and coworkers (Jacobs et al. 2015), the floor of a room would have to be covered by a salt gradient and a

Fig. 3.2 Olfactory navigation. Humans have been shown to be able to navigate using olfactory cues. (**a**) Subjects can follow a scent trail of chocolate essential oil (in *white*; track of subject in *black* (Porter et al. 2007). (**b**) In a room with two odor sources, subjects can olfactorily determine their position by analyzing the ratio of the two smells (Jacobs et al. 2015)

perpendicular sugar gradient. In this situation, each location in the room can be gustatorily identified by the taste intensity and the salt:sugar ratio.

However, that a stimulus is used for navigation does not mean that the perception of the stimulus has spatial structure. Even the most simple and unstructured signal can be used for navigation. Imagine an individual that gets headaches when standing close to a power line. Let us assume that the headaches have no spatial structure and that there are no qualitative differences between two headaches, but that the intensity of the headache depends on the distance to the power line. This individual could use their headache to find a power line and to follow it from one city to another using electricity-guided navigation. If, while walking in a straight line, the headache decreases, the individual knows that they are moving away from the power line. The ability to use the taste of salt or an electricity-induced headache for navigation does not show that the salt perception or the headache has a spatial structure. Any perception of a stimulus can be used for navigation when the stimulus has spatial structure.

The lack of spatial arrangements of perceptual qualities in olfaction is unusual. In vision, touch, and taste, sensory neurons are arranged spatially and their patterns of activation are mapped onto perception so that the spatial structure of the percepts resembles the spatial structure of the perceived stimulus. In vision and touch, this spatial mapping is central to the function of these senses. In gustation, the mapping does not seem to be central to the function of taste perception. However, although this information is rarely important, it is easy to tell the difference between a few grains of salt in different positions on the tongue. In audition, the main strategy for structuring perception spatially is to compare input to the two ears. This comparison makes it easy to auditorily distinguish a car approaching from the left from a car approaching from the right.

Unlike perception in these modalities, olfactory perception, as I have shown above, has no spatial structure. This suggests that in olfaction, the nature of the perceptual qualities rather than their spatial arrangement is the important information. The relative importance of the nature of perceptual qualities in olfaction is also reflected by the very large number of discriminable smells discussed in Chap. 1. A comparison between olfaction and vision reveals that there are fewer discriminable colors than smells. However, the many possible spatial combinations of colors in visual perception make vision an extraordinarily powerful modality.[4] To illustrate the relative importance of spatial structure in the two different modalities, imagine not seeing colors, but only grayscales and then compare it to the olfactory equivalent, smelling only different levels of odor intensity without distinguishable odor qualities. Colorless vision would still be a very useful sense. Many visually guided behaviors—manipulating tools, hunting boars, fighting—would be little affected by such a change because they rely on spatial information rather than on accurate information about colors. In contrast, odorless smell would be of extremely limited utility.

The situation found in olfaction, in which the type of perceptual quality rather than how perceptual qualities are arranged in space and time, is presumably the evolutionary oldest and most common situation. When it is cold, we know to wear a coat or show other heat preserving behaviors, regardless of the spatial and temporal structure of the cold. When we taste sugar, we should swallow what is in our mouth because it likely has high caloric content. When we taste bitter, we know to spit it out. The spatial and temporal structure of the gustatory perception is not important for guiding this behavior. In contrast, human visual perception evolved into a form of perception in which the arrangement of perceptual qualities in space has become the most important aspect. The differences between modalities in the importance of the nature and the spatial arrangements of perceptual qualities are a difference in degree[5]; however, these differences are the cause of how we tend to think about a given modality. They are the reason why our thinking about perception becomes distorted when it is based exclusively on one modality.

3.2 Is Olfactory Perception the Perception of Objects?

The most prominent distortion in the philosophy of perception is that it generalizes the importance of the spatial arrangement of perceptual qualities from vision to all of perception instead of treating it as a special adaptation in how humans perceive reflected light. This has led to several proposals about perception that are intuitive for visual perception, but do not generalize to other modalities or forms of perception. One of these proposals is that we do not perceive perceptual qualities, but perceptual objects.

If a given instance of perception can be either the perception of objects or objectless perception, then criteria to decide whether a given instance of perception involves objects or not has to be developed. I will discuss in this section the two criteria that are most frequently applied. The first criterion, which dominates the scientific literature, is that perception is the perception of objects when it involves figure-ground segregation (Kubovy and Van Valkenburg 2001, p. 102; Stevenson and Wilson 2007, p. 1823). The second criterion for perceptual objecthood that will be discussed in this section is that perception involves perceptual objects when the Many Properties Problem can be solved (Batty 2014; Carvalho 2014).The Many Properties Problem is the problem of distinguishing stimuli in which the same properties are instantiated in different arrangements. That we can tell the difference between a painting of a blue triangle next to a red square and a painting of a red triangle next to a blue square shows that vision can solve the Many Properties Problem.

Figure-Ground Segregation

Seeing a person standing in front of a wall or a banana lying on top of a table are instances of perceiving a figure (the person or the banana) that can be distinguished from the ground (the wall or table). For smell, whether percepts have a figure-ground structure is less obvious than for instances of visual perception. It has been argued that there is figure-ground segregation, and therefore objects of perception in olfaction (Stevenson and Wilson 2007). I will investigate this claim here. I will consider whether there is figure-ground segregation in olfaction first for olfactory perception without temporal structure and then for instances of olfactory perception in which the potential figure has a temporal structure.

At any given moment, a mixture of different odor molecules at different ratios activates our olfactory system. This ratio carries no spatial information (see Sect. 3.1.). A rose under your nose will produce the exact same activity of the olfactory sensory neurons as an intricate spatial arrangement of 275 vials, each of which containing the appropriate concentration of one of the 275 chemicals that are found in rose oil (Ohloff 1994, pp. 154–158). An apple under your nose in a vineyard can produce the exact same pattern of activity in the olfactory sensory neurons as a bunch of grapes under your nose in an orchard. More generally, a weak source of odor A that is placed close to the perceiver in front of a strong background of odor B can be indiscriminable from a weak source of odor B close to the perceiver in front of a strong background of odor A. Unlike visual perception, olfactory perception does not have a figure-background structure that reflects the relative positions of stimuli in physical space.

The conclusion that there is no figure-ground segregation in olfaction because olfactory percepts do not reflect positions of stimuli in physical space can be challenged in several ways. First, instead of being based on the perceived spatial relation between figure and ground, the distinction between figure and ground in olfaction could be based on intensity. Smells in nature are usually mixtures of many different odorants. When one of these odorants within the mixture dominates the perceived smell, it could metaphorically be thought of as "rising above the background". This could be interpreted as intensity-based figure-ground segregation. Imagine smelling a wine that smells strongly of apple as well as weakly of pineapple, cinnamon, and cherry. Smelling this wine could be described as a figure of apple with a background of pineapple, cinnamon, and cherry. Another way in which olfaction could accomplish figure-ground segregation is through differential familiarity. An odor mixture could acquire a figure-ground structure when its components differ in their familiarity. Imagine smelling a wine that has a weak apple smell as well as weak cardamom, turmeric, and elderberry smells. Imagine further that you never before smelled cardamom, turmeric, and elderberry. This situation can be described as perceiving an apple object and an unidentifiable background. Finally, a similar phenomenology can emerge due to violation of expectation. A wine that smells like a normal wine but also has an unexpected fishy smell can be described as a fish object and a wine background. In all these examples of olfactory perceptions that could be considered as having a figure-background structure, the distinction between figure and background is based on the prominence of some aspect of the olfactory percept. The prominence can be conveyed through intensity, familiarity, or unexpectedness. The prominent aspect is interpreted as the figure while the less prominent aspects form the background.

To decide if any of these situations should be considered as evidence for figure-ground segregation, it is informative to compare them to analogous situations in visual perception. The first case, strong apple odor perceived simultaneously with weaker other odors, corresponds to the visual scenery of looking at a tiled bathroom wall with differently colored tiles. The color of one tile is darker and more saturated than the colors of the other tiles. The second situation, apple odor together with an unknown mix of odors, corresponds to the visual experience of seeing a pile of unfamiliar-looking alien artifacts with a dead cow in between them. The third situation, unexpected fish odor mixed with regular wine odors, corresponds to the visual perception of a pink banana in a bunch of yellow bananas. A review of the literature on visual perception shows that such situations would be described as allocation of spatial attention depending on stimulus salience. They would not be considered instances of figure-ground segregation. Since salience-based attention allocation is not considered evidence for figure-ground segregation in vision, it should also not count as evidence for figure-ground segregation in olfaction.

My discussion of figure-ground segregation in olfaction so far has abstracted away from the temporal structure of perception. I have only shown that stationary olfactory snapshots do not have a figure-ground structure. Differential prominence of different features of an olfactory percept does not amount to figure-ground segregation. However, olfactory perception does also have a temporal structure. The figure-ground segregation in olfaction could be based on the fact that different sets of features of an olfactory percept covary over time (Wilson and Stevenson 2006; Carvalho 2014). This is an intriguing idea because our sense of smell is optimized to detect changes rather than states. We adapt quickly to any temporally constant odor and therefore stay sensitive to changes (Dalton 2000; Stevenson and Wilson 2007). The constant odor to which we are adapted, and therefore perceive only weakly, can be considered "background odor". Any change "in front" of this background can then be considered an object. Olfactory objects, according to this description, are similar to auditory events. There may be an olfactory background consisting of a mix of body odors and low-grade pollution and then a strawberry odor appears and lasts for a few moments before it disappears again. The strawberry odor that changed while the other odors remained constant can be thought of as an olfactory perceptual object.

As with cases in which certain aspects of a smell are more prominent, it is informative to consider whether covariation over time would be considered as evidence for figure-ground segregation in visual perception. Consider the banana as a figure and the table as a background. Now imagine that there is some localized damage and skin abrasion to the banana. Over time, the damaged spots of the banana will turn brown while the rest of the banana and the table do not change color. This would not result in a reassessment of the figure-ground segregation so that the brown spots on the banana are now considered the object and the yellow part of the banana and the table the background. Temporal changes do not introduce figure-ground relations in vision. It is therefore parsimonious that such temporal changes also do not introduce figure-ground relations in other modalities.

Another reason to resist a notion of perceptual objecthood that is based on the temporal structure of perception is that such a notion would be so inclusive that "perceptual object" would become a synonym of "percept". If appearance and disappearance of a perception are sufficient to make it the perception of an object, then all perceptual systems that detect change over time detect perceptual objects. A hypothetical sensory system that consists of a single sensory neuron can distinguish objects from background, when objects are defined by their temporal structure. The neuron will permanently be activated at a low level due to background levels of the stimulus. Occasionally, it will be strongly activated for a certain period of time when it encounters high levels of the stimulus. Afterward it will go back to the low level of activation. The periods of strong activation can be interpreted as the objects that can be segregated from the constant background. Under this notion of perceptual object, all perception is the perception of objects and the dispute about objecthood becomes a dispute about terminology.

The Many Properties Problem

Instead of using figure-ground segregation as the criterion of objecthood in perception, it has been suggested that perception is only the perception of objects when it can solve the Many Properties Problem. The Many Properties Problem (Jackson 1977) is the problem of assigning properties to objects in situations in which several objects are perceived simultaneously. When we see a blue circle and a red square, we can solve the Many Properties Problem and assign the properties "blue" and "round" to one object and the properties "red" and "square" to the other object. The way the Many Properties Problem is solved in visual perception is by grouping properties together based on their location in perceived space. Properties that move through space together are properties of the same object. In olfaction, there is no perceived space. Regardless, intuition suggests that different smells can be perceived at the same time. We can simultaneously perceive wine smell and sausage smell. If either the sausage smell or the wine smell has a spicy quality, can we assign the spiciness to one of the two smells? If we can, then olfaction can solve the Many Properties Problem.

It has been suggested that olfaction can solve the Many Properties Problem and that "spicy wine, and sausage" does in fact smell different from "wine, and spicy sausage" (Carvalho 2014, p. 12). However, no biological mechanism on which this capacity could be based has been suggested. The molecules that arrive at the olfactory epithelium and elicit the percept of "spicy wine and sausage" can be identical to the molecules that elicit the percept of "wine and spicy sausage". One of the many molecules that are perceived as spicy is cuminaldehyde. Cuminaldehyde is found in some wines and it gives the smell of those wines a spicy note. However, cuminaldehyde is also given off by some spicy sausages. When the cuminaldehyde molecules reach the olfactory sensory neurons, they do not carry information about their source. The perception is the same regardless of whether the cuminaldehyde is given off by the sausage or by the wine. Olfaction cannot solve the Many Properties Problem (Batty 2010, 2011). Because we cannot tell which property belongs to which potential olfactory object, we also cannot tell how many potential olfactory objects contribute to a given instance of perception. Instead of smelling spicy sausage and wine, or spicy wine and sausage, we may smell wine, sausage, and a jar of cumin oil.

The finding that we cannot assign properties to smells seems to defy everyday experience. When an anchovy and garlic pizza is delivered, we know that the anchovy smell and the garlic smell are both properties of the pizza smell. However, this knowledge is based on background knowledge and assumptions rather than on perceptual grouping of the anchovy and garlic smell. If the pizza that is delivered does not have anchovies on it, but the pizza deliveryman carries some anchovies in his pocket to snack on, we will incorrectly assume that both the garlic smell and the anchovy smell are properties of the pizza smell. This is because our assigning of properties to smells is not based on the structure of our percepts, but on our assumptions about the most likely scenarios. Similarly, somebody who is familiar with spicy sausages but not with spicy wine will interpret the stimulus discussed above as the perception of wine and spicy sausage. Somebody familiar with spiced wines, but not with spiced sausages will interpret the same stimulus as the perception of spiced wine and sausage. Nothing in the olfactory percept itself enables us to assign the spiciness to either of the two smells.

Just like figure-ground segregation, solving the Many Properties Problem is possible in olfaction when the temporal structure of olfactory perception is taken into account. Covariance between the sausage and spice smell over time is good evidence that they are released by the same odor source and that they are properties of the same object. However, as discussed above, if temporal criteria would be sufficient for objecthood, then all perception with temporal structure would be the perception of objects and "object-based perception" would just be a different label for "perception".

3.3 What Could Be an Olfactory Object?

In Sect. 3.2., I used two different criteria for objecthood to investigate whether olfactory perception involves the perception of objects. The result of this investigation was that the only conceptual framework under which olfactory perception is the perception of objects is one under which every instance of perception is the perception of objects. If such a definition of objecthood is endorsed, the question whether olfactory perception is the perception of objects is not an empirical question. Instead, olfactory perception is by definition the perception of olfactory objects. In this section, I will investigate what the objects of olfactory perception that are postulated by such an approach could be.

I will first discuss candidate olfactory objects that have been suggested in the literature. The most obvious candidates for olfactory objects are the source of the odorous molecules, the cloud of odorous molecules that the source gives off, and the molecules themselves. I will show that neither of these candidates have the properties usually expected from objects of perception. I will also discuss a more flexible notion of perception that uses phenomenal presence as the criterion for objecthood. This criterion has been suggested by Budek and Farkas, who propose that the causes of an instance of perception that are present in phenomenology are the objects of that instance of perception (Budek and Farkas 2014).

Potential Odor Objects

A straightforward proposal about olfactory objects is that the source of the odorous molecules is the object of the olfactory perception. Benjamin Young calls this view, which is based on naive realism about perception (Gibson 1966), the "ordinary object view" (Young 2011, p. 46). According to the ordinary object view, the objects of the smell of lions are lions. One problem with this view is that the source of odorous molecules can be specified at different levels. What appears to be lion smell is mostly the smell of lion urine, so both the urine and the lions themselves are possible objects of the perception. Somebody driving by the zoo may think of the entire zoo as the object of the lion smell. Lion urine, individual lions, the pack of lions, the lion enclosure, and the zoo all can be considered as the source of the lion smell and therefore as its object.

Another problem of the ordinary object view of olfactory perception is that the link between the source of molecules that cause olfactory perception and the perception is much weaker and less direct than the link between the source of the light that causes visual perception and the corresponding visual perception. The smell of lions will linger for a long time after the lions have been moved to another zoo. The lion odor will also stick to their zookeeper's clothes so that when she comes home, she will smell of lions. Are the zookeeper's clothes now the object of perception? These examples show that smells, after they are given off by their sources, are no longer linked to them. It is therefore difficult to identify the "original" source of a smell.[6]

Because of these problems with applying the ordinary object view to olfaction, several alternatives have been proposed. One alternative suggestion is that the cloud of odorous molecules given off by the odor source is the object of olfactory perception (Lycan 2000). This cloud object view avoids the problems of unambiguously identifying the source of the cloud of molecules that make the ordinary object view unattractive for olfaction. The disadvantage of the cloud object view is that it cannot be integrated into an evolutionary account of perception. The perception of lions provides useful information that can guide behaviors that increase the perceiver's fitness by decreasing the likelihood of being eaten by a lion. The perception of a cloud of odor molecules, on the other hand, is in most cases not adaptive (as has been pointed out by Ruth Millikan in conversation with Bill Lycan (2014a, p. 7)).

Another problem with the cloud object view is that an odor source usually gives off hundreds of different odorous molecules. Presumably, under the cloud object view, each type of molecule is considered to form its own cloud. The perception of the smell of lions therefore is the perception of hundreds of different objects. Alternatively, all the different types of molecules given off by the lions could be considered to form a single cloud. However, at any time, different types of molecules that have been given off by different odor sources are in the air. The only thing the different types of molecules that are given off by a lion have in common is that they have been released from the same source. The only way to determine which of the many molecules in the air belong to the lion odor cloud is therefore by reference to the odor source. This reintroduces all the problems of the ordinary object view that the cloud object view was supposed to solve.

An alternative to the ordinary object view and the cloud object view is the molecule object view, according to which the odorous molecules that bind to our odorant receptors are the olfactory objects. Benjamin Young has developed a proposal along these lines, although he does not consider the entire odorous molecule, but just its chemical structure, to be the object of olfaction (Young 2011). The advantage of a molecule object view over the cloud object view and the ordinary object view is that the perceptual qualities associated with olfactory perception are related to properties of the molecules, not to properties of the ordinary object or the cloud. Depending on their chemical and physical properties, different odor molecules interact with different combinations of odorant receptors. This combinatorial code determines the perceived smell. Differences in molecules therefore result in differences in perception. With the molecule object view, the correlation between physical differences between two objects and differences in the perceptual qualities associated with the two objects is much higher than with the ordinary object view or the cloud object view.

Accounting for the differences and similarities in perceived smells is an important advantage of the molecule object view. However, the molecule object view faces the same problems faced by the cloud object view. It fails to provide a connection between perception and adaptive behavior. Smelling macrocyclic rings with two hydrogen bonds at the central oxygen conveys no adaptive advantage. Another problem that the molecule object view inherits from the cloud object view is the unsatisfying account of the perception of odorant mixtures. Wine gives off hundreds of different types of molecules that have different chemical structures (Aznar et al. 2001). If molecules (or their structure) are the objects of olfactory perception, then we perceived hundreds of different objects when opening a bottle of Chardonnay.

The ordinary object view, the cloud object view, and the molecule object view all have their advantages and disadvantages. Maybe the advantages of all three views could be combined. The zoo, the lions, their urine, the cloud of molecules that forms above the urine, and the structure of the molecules in that cloud could *all* be objects of olfaction. Such a layered account of objecthood has been suggested by Lycan (2014a, b). The layered account of objecthood is inspired by the layered accounts of referring. Lycan's example of a layered account of referring is that we can point at a chalk mark on a board and thereby refer to a number; thereby we refer to a room; thereby we refer to the occupant of that room. The analogous account for olfaction would be that by smelling a certain molecule, we smell the cloud of molecules it belongs to; thereby smelling the lions that gave off the cloud of molecules; thereby smelling the zoo in which the lions are housed. This account does combine the advantages of the three accounts discussed above, but it also combines their disadvantages.

Phenomenal Presence as a Criterion for Objecthood

As we have seen, different authors have suggested different objects of olfactory perception. The competition between the different proposals focuses on weighing the advantages and disadvantages of one proposal against the advantages and disadvantages of competing proposals. It is a disadvantage of the molecule object view that according to it the perception of a Chardonnay is the perception of hundreds of different objects. On the other side, the molecule object view has the advantage that it accounts partially for the differences and similarities in perceived smells. Whether the advantages of any of the proposed views are worth accepting the counterintuitive consequences that come with it is a matter of preferences. An alternative procedure to decide between the different proposals is to apply a criterion for objecthood. One proposal for identifying the object of an instance of perception is to use phenomenal presence as the criterion for objecthood. Following this proposal, the objects of an instance of perception are the causes of that instance of perception that are present in phenomenology (Budek and Farkas 2014).[7]

Every instance of perception has causes. The perception of rose odor is caused by the rose and by the clouds of odor molecules given off by the rose and by the binding of those molecules to olfactory sensory neurons. One could consider all these causes of the instance of perception to be its objects. Lycan's layered account of objecthood suggests something similar. However, calling the causes of an instance of perception its objects does not result in an interesting notion of objecthood. Instead, it merely results in a change in terminology from "causes" to "objects". The proposal suggested by Budek and Farkas is that only a subset of the causes of perception, namely those causes that are phenomenally present, are the objects of perception (Budek and Farkas 2014). We may "see" that it is cold outside because it is snowing, but the temperature is not present in our visual phenomenology and therefore not an object of our visual perception. The snow, on the other hand, is present in our phenomenology and is therefore an object of this instance of perception. The objects are whatever is phenomenally present in a given instance of perception.

The methods of determining what is phenomenally present are problematic because the main evidence for phenomenal presence is introspection, and introspections differ between individuals. This variability has resulted in different proposals about the objects of visual perception. Some theorists have argued that in instances of visual perception only colors and shapes are present in phenomenology (e.g., Tye 1995). Others have reported that objects such as pine trees are present (e.g., Siegel 2006). I will now discuss which of the potential objects of olfactory perception discussed above are phenomenally present in instances of olfactory perception.

Molecules, their structures, or clouds of molecules are not normally present in phenomenology. That smell is mediated by molecules that form clouds (instead of, e.g., by waves of a certain wavelength) is not self-evident but had to be discovered. Many people who have smell experiences are not aware that they are experiencing clouds of molecules. It does not seem that their olfactory perceptions change once they are informed that the perceptions are caused by molecules. In some cases, molecules are present in phenomenology, though. Consider, for example, a fragrance chemist who is synthesizing new odor molecules. When the chemist smells a new molecule for the first time after synthesizing it, the molecule, or maybe the molecule's structure, will be present in phenomenology. The synthetic molecule, which is not found in nature, does not have an ordinary object that gives off the molecule. Alternatively, one could say that the ordinary object of the synthetic molecule is a liquid-filled test tube with the molecule's name written on it. Let us consider a second example. The janitor of a building is attempting to locate the source of an unidentified malodor. The janitor does not know what the object giving off the odor is. What is present in phenomenology for the janitor is the odor cloud and how it travels through the building. After the source is identified as a dead rat, the dead rat, as the ordinary object, is present in phenomenology. These considerations show that what is present in phenomenology in olfaction depends on the background knowledge of the perceiver. It can be molecules, clouds of molecules, objects that give off odors, or combinations of the three. The same stimulus can be veridically perceived by perceivers with different background knowledge, yet what is present in phenomenology in the different perceivers, and therefore the object of perception, differs.[8]

3.4 Conclusion: Olfactory Perception Is Not the Perception of Objects

I do not deny that most instances of perception appear to us as if they are the perception of objects. Consider, for example, the tactile perception of an object that is stuck in your throat, or under your eyelid. The object that is stuck under the eyelid has a certain size and it is located at a certain position. The object also has non-spatial properties. It may be hot or rough, for example. As we know, the perception of an object stuck under the eyelid is in many cases illusionary. Such a perception is often not caused by a physical object. Instead, the perception of something being stuck in the perceiver's throat or under her eyelid is usually caused by localized irritation of the skin or mucosa. This illusion of objecthood shares many features with other types of illusions. The illusion of objecthood, for example, resists cognitive influence. Even when my doctor inspects the inside of my eyelid and informs me that the percept is caused by localized irritation and not by a physical object, the illusion of perceiving an object under my eyelid continues.

A thought experiment shows how tightly illusory objecthood is linked to the spatial arrangement of perceptual qualities. Imagine that the skin irritation under the eyelid starts to spread. Once the perceptual qualities have spread to cover half of the perceiver's face, they are not perceived as an object that covers a large area of the face. Instead, they are now perceived as spatially arranged perceptual qualities. Because the perception of perceptual qualities that are spatially localized has a strong tendency to induce the perception of illusory objects, olfaction, in which perceptual qualities are not arranged spatially (discussed in Sect. 3.1.), is an ideal system to investigate perceptual objecthood without this complication.

I have discussed two ways of approaching the question of perceptual objects. One approach (discussed in Sect. 3.2.) is to establish criteria to decide when an instance of perception involves objects and then investigate whether a given instance of perception is the perception of an object or not. The alternative approach (discussed in Sect. 3.3.) is to postulate that all perception is the perception of objects. With this approach, the question is not whether perception involves perceptual objects, but whether perception is best *described* as the perception of objects.

Concerning the first approach, I investigated two criteria that allow for an empirical test whether an instance of perception involves objects or not. The two criteria for objecthood that I discussed are susceptibility to figure-ground segregation and capacity to solve the Many Property Problem. I argued that olfactory perception fails to satisfy both criteria.[9] I have shown that in cases in which it has been argued that olfactory perception satisfies a criterion for objecthood (Wilson and Stevenson 2006; Carvalho 2014), the criterion for objecthood is so broad that it is satisfied by all temporally structured instances of perception. Because every perception has a beginning and an end, this results in the conclusion that all perception is the perception of objects.

Empirical testing for objecthood can be dispensed off when all perception is postulated to be perception of objects. There are many notions of objecthood that are so loose that they include all instances of olfactory perception. Olofsson, for example, writes that the unified olfactory percept is "commonly referred to as an olfactory object" (Olofsson 2014, p. 2). This approach suggests that there is no difference between the perception of objects and other perception, but that it is preferable, for theoretical reasons, to think of all instances of perception as involving objects. The important question, when all perception is postulated to involve objects, is what those objects are in the case of olfaction.

I have reviewed the advantages and disadvantages of some suggested objects of olfactory perception: the odor source, clouds of odor molecules, and odor molecules or their structure. I then used phenomenal presence as the criterion to decide what the object of olfactory perception is.

The discussion revealed that what is phenomenally present in olfactory perception depends largely on the perceiver's background knowledge. When a perfumer's olfactory sensory neurons are activated by the synthetic molecule Galaxolide, the Galaxolide molecule, or its structure, is phenomenally present for the perceiver. When the same molecule activates the olfactory sensory neurons of an average consumer, laundry detergent, or fresh laundry is phenomenally present. Galaxolide is added to laundry detergent to give it its characteristic smell. When somebody living in a part of the world in which laundry detergent is not common perceives Galaxolide, presumably what is present in phenomenology is just some sort of smelliness.

"Some sort of smelliness" is not an object. This, I suggest, is because olfaction is not an object-directed process. It is therefore not justified to postulate that all perception is the perception of objects. The notion of objects unnecessarily complicates the description of olfactory perception. An alternative view, which I endorse, has been proposed by Clare Batty, who holds that what we perceive in olfaction is not objects, but existentially quantified properties ("There is F-ness here") (Batty 2010, 2011). According to Batty, olfactory perception does not involve objects that have properties. Instead, in olfaction, perceptual properties are "free-floating" or "object-less" (Matthen 2005; Batty 2010). Batty calls this view, which is motivated by many of the same considerations discussed in this chapter, the *abstract view* of olfactory content.

Notes

1. As reasons why it is justified to think of audition as involving objects, it has been pointed out that "audible individuals are temporally extended and bounded, serve as the locus for auditory attention, prompt completion effects, and are subject to figure-ground distinctions in pitch space" O'Callaghan (2014). "Auditory Perception." *The Stanford Encyclopedia of Philosophy*, from http://plato.stanford.edu/archives/sum2014/entries/perception-auditory/. O'Callaghan (2008). "Object perception: Vision and audition." *Philosophy Compass* **3**: 803–829 proposed that objects in vision and in audition are mereologically complex individuals. In vision, complexity is conveyed by the perception's spatial features, whereas in audition temporal structure makes perception complex.

2. For a defense of olfactory objects, see Stevenson (2014). "Object concepts in the chemical senses." *Cognitive Science* **38**(7): 1360–1383.
3. Chemesthesis is the chemical sensitivity of the skin or mucous membrane. The hotness when eating chili peppers, but also the burning in the eye when fingers touch the eye after handling chili peppers are mediated by chemesthesis. The sting when smelling vinegar and the perceived coolness of menthol when it is inhaled are other examples of chemesthetic perception. Chemesthesis is mediated through receptors that mediate pain, touch, and temperature, which explains why chemesthetic experiences are described in terms of temperature and touch perception.
4. To illustrate just how many different visual percepts humans can discriminate, imagine a chess board with different combinations of the black and white pieces on them. There are around 10^{71} such combinations. I suspect, although it should be tested, that it is possible to tell apart any of these combinations, including those that have a minimal difference, for example, a black pawn being on E6 rather than E7. If that is the case, then humans can distinguish more than a quantazillion (one trillion times one trillion times one trillion times one trillion) visual percepts. And of course we can not only distinguish black and white chess pieces, but also red, green, and blue ones, which increases the number of possible combinations exponentially. Furthermore, humans can distinguish different faces and trees from chess boards with different combinations of pieces. The actual number of different visual stimuli that can be discriminated thanks to the different spatial combinations in which colors can be arranged is incomprehensibly large.
5. All perception needs discriminable perceptual qualities to be adaptive. Spatial and temporal arrangements of perceptual qualities are only possible when there are at least two different perceptual qualities.
6. How strongly the messengers of perception are linked to their source is a matter of degree. The difficulties pointed out here are not unique to olfaction; they are, however, much more pronounced in olfaction than in other modalities. An example of a weak link between the messengers of perception and their source in vision is our visual perception of objects that are very far away. We can see stars that no longer exist, because light from distant stars can take a very long time to reach our eyes and therefore sometimes a star will cease to exist before its light reached us. Even the stars that still exist are perceived to be at the position in the sky where they were when they gave off the light and not where they are when we perceive them. The path of light can also be changed through gravitational deflection. All these phenomena make the link between light and its source less direct. In most cases of everyday perception,

the connection between light and its source is, however, much more direct than the connection between an odor and its source.

7. According to the proposal by Budek and Farkas, every instance of conscious perception is an instance of the perception of objects because in every instance of conscious perception *something* is phenomenally present. The criterion of phenomenal presence is different from the criteria discussed above (figure-ground segregation; Many Properties Problem) in that it is not used to decide *whether* an instance of perception involves objects or not. Instead, the criterion of presence in phenomenology presupposes that instances of conscious perception involve objects. It is used to decide *what* these objects are.
8. That the nature of olfactory objects depends on the perceiver has also been suggested based on different considerations. Yeshurun and Sobel write:

   We suggest that an odor object is the integration of the odor's inherent pleasantness (...) with the subjective state at the moment of coding: mood, hunger, fear, etc. Therefore, an odor object is not the odor of the banana but rather an integration of the pleasantness of the banana odor with the subjective state at which it was encountered. Yeshurun and Sobel (2010). "An odor is not worth a thousand words: From multidimensional odors to unidimensional odor objects." *Annual Review of Psychology* **61**(1): 219–241. (p. 229)

   In this context, what the object of olfactory perception is does not depend on the perceiver's background knowledge, but on the perceiver's attitude toward the odor and on their current behavioral goals. A similar proposal suggests that the object of olfactory perception is the biological value to the perceiving organism. Castro and Seeley (2014). "Olfaction, valuation, and action: Reorienting perception." *Frontiers in Psychology* **5**. What is phenomenally present when an odor is perceived, according to these proposals, is the pleasantness, or the value, of the odor. The pleasantness, or value, is partially caused by the perceiver's state. Food odor is more pleasant for a hungry perceiver. However, it does not seem that hunger is actually phenomenally present in the perception of food odor when hungry. Instead, this situation seems to be similar to the role of cold in perceiving snow. The outside temperature changes the visual perception of water falling from the sky from rain to snow, but despite this influence on the visual perception, the cold itself is not phenomenally present in vision.
9. That olfactory perception in humans does not satisfy these criteria for objecthood is contingent on the structure of our olfactory system. If our body were covered in odor-sensitive cells, then some instances of olfactory perception

could have a spatial structure and we could sometimes be able to distinguish between an odor figure and a background. Under these circumstances, olfaction's ability for figure-ground segregation would be similar to figure-ground segregation in thermoperception. We can segregate the sun's heat radiation from background radiation. The sun is then the object of our perception of heat. Similarly, a hot potato that we hold in our hand can easily be discriminated from the temperature background.

References

Aznar, M., López, R., et al. (2001). Identification and quantification of impact odorants of aged red wines from Rioja. GC-olfactometry, quantitative GC-MS, and odor evaluation of HPLC fractions. *Journal of Agricultural and Food Chemistry, 49*(6), 2924–2929.

Barwich, A.-S. (2014). A sense so rare: Measuring olfactory experiences and making a case for a process perspective on sensory perception. *Biological Theory, 9*(3), 258–268.

Batty, C. (2010). A representational account of olfactory experience. *Canadian Journal of Philosophy, 40*(4), 511–538.

Batty, C. (2011). Smelling lessons. *Philosophical Studies, 153*, 161–174.

Batty, C. (2014). The illusion confusion. *Frontiers in Psychology, 5*, 1–11.

Bizley, J. K., & Cohen, Y. E. (2013). The what, where and how of auditory-object perception. *Nature Reviews Neuroscience, 14*(10), 693–707.

Blauert, J. (1997). *Spatial hearing: The psychophysics of human sound localization.* Cambridge: MIT Press.

Budek, T., & Farkas, K. (2014). Which causes of an experience are also objects of the experience? In B. Brogaard (Ed.), *Does perception have content?* (pp. 351–369). Oxford: Oxford University Press.

Carvalho, F. (2014). Olfactory objects. *Disputatio, 6*(38), 45–66.

Castro, J. B., & Seeley, W. P. (2014). Olfaction, valuation, and action: Reorienting perception. *Frontiers in Psychology, 5*,(299), 1–4.

Classen, C., Howes, D., et al. (1994). *Aroma: The cultural history of smell.* London: Routledge.

Croy, I., Schulz, M., et al. (2014). Human olfactory lateralization requires trigeminal activation. *NeuroImage, 98*, 289–295.

Dalton, P. (2000). Psychophysical and behavioral characteristics of olfactory adaptation. *Chemical Senses, 25*(4), 487–492.

Frasnelli, J., Charbonneau, G., et al. (2008). Odor localization and sniffing. *Chemical Senses, 34*(2), 139–144.

Gagliardo, A. (2013). Forty years of olfactory navigation in birds. *Journal of Experimental Biology, 216*(12), 2165–2171.
Gibson, J. J. (1966). *The senses considered as perceptual systems*. Boston: Houghton Mifflin.
Griffiths, T. D., & Warren, J. D. (2004). What is an auditory object? *Nature Reviews Neuroscience, 5*(11), 887–892.
Heidegger, M. (2002/1950). The origin of the work of art. In J. Young and K. Haynes (Eds.), *Off the beaten track*. Cambridge: Cambridge University Press.
Jackson, F. (1977). *Perception*. Cambridge: Cambridge University Press.
Jacobs, L. F., Arter, J., et al. (2015). Olfactory orientation and navigation in humans. *PLoS One, 10*(6), e0129387.
Kleemann, A. M., Albrecht, J., et al. (2009). Trigeminal perception is necessary to localize odors. *Physiology & Behavior, 97*(3–4), 401–405.
Köster, E. P. (2002). The specific characteristics of the sense of smell. In C. Rouby, B. Schaal, D. Dubois, R. Gervais, & A. Holley (Eds.), *Olfaction, taste, and cognition* (pp. 27–44). Cambridge: Cambridge University Press.
Köster, E. P., Møller, P., et al. (2014). A "Misfit" theory of spontaneous conscious odor perception (MITSCOP): Reflections on the role and function of odor memory in everyday life. *Frontiers in Psychology, 5*, 64.
Kubovy, M., & Van Valkenburg, D. (2001). Auditory and visual objects. *Cognition, 80*(1–2), 97–126.
Lycan, W. G. (2000). The slighting of smell. In N. Bhushan & S. Rosenfeld (Eds.), *Of minds and molecules: New philosophical perspectives on chemistry* (pp. 273–289). Oxford: Oxford University Press.
Lycan, W. G. (2014a). The intentionality of smell. *Frontiers in Psychology, 5*, 436.
Lycan, W. G. (2014b). What does vision represent? In B. Brogaard (Ed.), *Does perception have content?* (pp. 311–328). Oxford: Oxford University Press.
Matthen, M. (2005). *Seeing, doing, and knowing*. Oxford: Oxford University Press.
Matthen, M. (2010). On the diversity of auditory objects. *Review of Philosophy and Psychology, 1*, 63–89.
Nudds, M. (2010). What are auditory objects? *Review of Philosophy and Psychology, 1*, 105–122.
O'Callaghan, C. (2008). Object perception: Vision and audition. *Philosophy Compass, 3*, 803–829.
O'Callaghan, C. (2014). Auditory perception. *The stanford encyclopedia of philosophy*, from http://plato.stanford.edu/archives/sum2014/entries/perception-auditory/
Ohloff, G. (1994). *Scent and fragrances: The fascination of odors and their chemical perspectives*. Berlin: Springer-Verlag.

Olofsson, J. K. (2014). Time to smell: A cascade model of human olfactory perception based on response-time (RT) measurement. *Frontiers in Psychology, 5*, 33.

Porter, J., Anand, T., et al. (2005). Brain mechanisms for extracting spatial information from smell. *Neuron, 47*(4), 581–592.

Porter, J., Craven, B., et al. (2007). Mechanisms of scent-tracking in humans. *Nature Neuroscience, 10*(1), 27–29.

Radil, T., & Wysocki, C. J. (1998). Spatiotemporal masking in pure olfaction. *Annals of the New York Academy of Sciences, 855*, 641–644.

Siegel, S. (2006). Which properties are represented in perception? In T. Szabó Gendler & J. Hawthorne (Eds.), *Perceptual experience* (pp. 481–503). Oxford: Oxford University Press.

Stevenson, R. J. (2009). Phenomenal and access consciousness in olfaction. *Consciousness and Cognition, 18*(4), 1004–1017.

Stevenson, R. J. (2014). Object concepts in the chemical senses. *Cognitive Science, 38*(7), 1360–1383.

Stevenson, R. J., & Wilson, D. A. (2007). Odour perception: An object-recognition approach. *Perception, 36*, 1821–1833.

Tye, M. (1995). *Ten problems of consciousness*. Cambridge: MIT Press.

Varendi, H., & Porter, R. H. (2001). Breast odour as the only maternal stimulus elicits crawling towards the odour source. *Acta Paediatrica, 90*(4), 372–375.

Varendi, H., Porter, R. H., et al. (1994). Does the newborn baby find the nipple by smell. *Lancet, 344*(8928), 989–990.

von Békésy, G. (1964). Olfactory analogue to directional hearing. *Journal of Applied Physiology, 19*(3), 369.

Wallraff, H. G., & Andreae, M. O. (2000). Spatial gradients in ratios of atmospheric trace gases: A study stimulated by experiments on bird navigation. *Tellus Series B-Chemical and Physical Meteorology, 52*(4), 1138–1157.

Weissburg, M. J. (2000). The fluid dynamical context of chemosensory behavior. *Biological Bulletin, 198*(2), 188–202.

Wilson, D., & Stevenson, R. J. (2006). *Learning to smell: Olfactory perception from neurobiology to behavior*. Baltimore: The Johns Hopkins University Press.

Yeshurun, Y., & Sobel, N. (2010). An odor is not worth a thousand words: From multidimensional odors to unidimensional odor objects. *Annual Review of Psychology, 61*(1), 219–241.

Young, B. D. (2011). Olfaction: Smelling the content of consciousness. *Philosophy*. New York, City University of New York. PhD.

4

The Function of Perception

An important step to understanding any biological system is to identify its evolutionary function. What is the evolutionary function of perception?[1] Why have so many living things evolved elaborate systems to distinguish perceptual properties and to arrange them in spatial and temporal patterns? In this chapter, I will argue, perhaps uncontroversially, that it is the function of perception to guide behaviors. I will contrast this proposal with the alternative proposal that it is the function of perception to collect accurate information about the physical world.[2] In many instances of perception, this seems like a false dichotomy. Is it not possible that perception guides behaviors *by* collecting accurate information about the world? I will show here that in some situations guiding behaviors and collecting accurate information about the world are in conflict. In these situations, behavior guidance is *always* given priority over perceptual accuracy. I therefore propose that perception was shaped by natural selection to guide behaviors adaptively. The fact that in many cases it also collects accurate information about the physical world is due to the constraint that in some, or maybe even most, situations, behavioral guidance requires correct information about the physical features of the environment. The collection of accurate information through perceptual systems is not a direct product of evolution. Perception-dependent

A. Keller, *Philosophy of Olfactory Perception*,
DOI 10.1007/978-3-319-33645-9_4

behaviors are the direct products of evolution through natural selection; the collection of correct information is a spandrel.[3]

Let me illustrate the difference between correctness of information and behavioral guidance using subway maps as an example. The function of subway maps is to help people navigate the subway system. It is *not* the function of subway maps to provide accurate information about the physical location of subway lines and stations. Of course, in many cases subway maps do provide accurate geographical information. However, whenever the map designer has to make a decision between two designs, one of which is more accurate and the other more useful, she will decide to make the map as useful as possible for those trying to navigate the subway system. This is because it is the function of the subway map to help riders navigate the subway system. That is what the map is for.[4]

How can it be that reduced geographical accuracy of a subway map increases its usefulness as a tool for navigating the subway system? For the answer to this question, one has to compare a city's subway map with a geographically accurate map of the same city. Take New York City as an example. In Fig. 4.1a, a section of the official Metropolitan Transportation Authority subway map of New York City is shown. In Fig. 4.1b, the same section from an alternative, geographically more accurate, map is shown for comparison. The designers of the official subway map have sacrificed accuracy in the geographical information conveyed by the map to increase the map's usefulness in many instances. Most obviously, in the official map, Manhattan is represented as less narrow than it is. This makes it easier to discriminate and follow the many parallel train lines connecting uptown Manhattan to downtown Manhattan. The official subway map also smoothens the tracks of the lines, as can be seen, for example, in the section of the B and D train within the red rectangle in Fig. 4.1a and b. In the red circle at the southern tip of Manhattan, it can be seen how the designers of the official map distorted the map to

Fig. 4.1 Two ways of representing the subway stations and tracks in Manhattan. A section of the official MTA subway map of New York City (**a**) and the corresponding section of a geographically more accurate alternative subway map (**b**) are shown. The red rectangles and circles and the black rectangles mark corresponding locations on the two maps

a
b
Marble Hill 225 St
Kingsbridge Rd
Allerton Av
Pelham Pkwy
Fordham Rd
Bronx Park East
215 St
207 St
Inwood–207 St
183 St
Burnside Av
Tremont Av
Dyckman St
191 St
190 St
176 St
174 St
181 St
Mt Eden Av
170 St
Freeman St
175 St
167 St
Simpson St
168 St
Intervale Av
Prospect Av
Jackson Av
157 St
155 St
Harlem 148 St
145 St
3 Av–149 St
137 St City College
135 St
3 Av 138 St
125 St
116 St Columbia U
116 St
Cathedral Pkwy (110 St)
Central Park N (110 St)
110 St
103 St
96 St
MANHATTAN
86 St
81 St–Museum of Natural History
79 St
72 St
CENTRAL PARK
77 St
68 St Hunter College
66 St Lincoln Center
59 St Columbus Circle
57 St
Lexington Av/59 St
Queensboro Plaza
Lexington Av/53 St
50 St
49 St
47–50 Sts Rockefeller Ctr
51 St
Court House Sq
UNITED NATIONS
42 St Port Auth
Times Sq
42 St–5 Av Bryt Pk
Grand Central
34 St
Penn Sta
Herald Sq
33 St
28 St
23 St
18 St
14 St
8 St NYU
Astor Pl
Christopher St Sheridan Sq
W 4 St Wash Sq
Houston St
Prince St
Spring St
Canal St
Franklin St
Chambers St
City Hall
World Trade Center
Fulton St
Wall St
Cortlandt St
Rector St
Broad St
Bowling Green
South Ferry
Whitehall St
Hoyt–Schern
Hudson River
East River
225 St Marble Hill
Jerome Av Line
Metro-North
Broadway-7 Av Line
8 Av Line
Lenox Av Line
Lexington Av Line
Amtrak
Roosevelt Island
Amtrak/LIRR
42 St–LI City
Grand Central
Canarsie
William
PATH

separate stations that are very close to one another. This makes it easier to identify and distinguish these stations.

Another geographical inaccuracy of the official subway map that its designers have purposefully introduced to make it more useful is that they exaggerated the width of the Harlem River that separates Manhattan and the Bronx (black rectangles in Fig. 4.1a and b). How does representing the river to be wider than it is make the map more useful? The river is a barrier and people taking the subway should be made aware of this barrier so that they can plan to exit the subway at a station that is on the same side of the river as their destination. How wide the river is does not matter, what matters is that it constitutes a barrier. This is a typical situation in which geographical accuracy of a map and usefulness of a map are in conflict. Map designers always sacrifice representational accuracy for usefulness, although there are disputes about how far maps should be abstracted to maximize their usefulness (Jabbour 2010).

Just as it is the function of subway maps to help riders navigate the subway system, the function of perception is to guide the perceiver's behavior. Subway maps often, but not always, use accurate geographical representations to help riders navigate the subway system and perception often collects accurate information about the physical world to guide behavior. However, collecting accurate information about the physical world is only one of many different strategies used by perception to fulfill its function. Collecting accurate information is the *intermediate* function of *some forms* of perception, whereas guiding behaviors is the *ultimate* function of *all* perception.

Here, I will support this proposal about the function of perception by showing that chemosensory perception fails spectacularly at collecting accurate information about the physical world. By "accurate" or "correct" perception, I mean perception that reflects similarities in the physical world in similarities in perception. According to this definition, physically similar things are perceived correctly, when they are perceived to be similar. In Sect. 4.1., I will review evidence that different individuals, especially when they belong to different species, often perceive the same stimulus differently. The perceptions of a physical object of two different perceivers are usually not identically, which shows that at least one of the perceivers perceives the object incorrectly. In Sect. 4.2., I will then show that even within the same perceiver perception does not accurately reflect similarity relations in the physical world. In all the cases that I will discuss, perceivers have evolved to perceive the physical world less accurately than they could. In contrast, they perceive in all those cases the world in a way that ensures that they interact with it behaviorally in the most beneficial way for them. This discrepancy indicates that perception evolved to fulfill behavioral needs rather than to collect accurate information about the world. Those who resist a behavior-centered conception of the function of perception have responded to the type of evidence I will present in Sects. 4.1 and 4.2 by redefining "correctness". In Sect. 4.3, I will discuss the alternative notions of correctness that have been proposed and show that they result in a notion of correctness according to which perception is correct when it elicits an adaptive behavioral response. When such a behavior-based notion of correctness of perception is endorsed, then guiding behaviors and collecting correct information are the same thing and the conflicts between them that exist when correctness of perception is based on a relation between what is perceived and how it is perceived disappear.

4.1 Perceptual Variability

If all living things had evolved to collect accurate information about the physical world, one would expect all living things to collect very similar information, since they all live in the same physical world. In this section, I will show that this is not the case. I will provide a few examples of perceptual differences between different species, as well as examples of how members of the same species perceive the same stimulus differently, depending on their behavioral goals.

Perceptual Variability Between Species

Comparative ecology has shown that different animal species often differ dramatically in how they perceive the physical world. Often this variability can be explained by the ecological niche inhabited by the species in question and by the behavioral repertoire with which the species interacts with the environment. Bees, for example, visit flowers, where they collect nectar. It is therefore important for bees to be able to distinguish flowers. Since many flowers have UV patterns, bees evolved the ability to detect

Fig. 4.2 Color perception in different animal species. (**a**) The nectar guide, a UV pattern that guides pollinators to the nectar, is invisible in this *Mimulus* flower to humans (*left*), but it can be perceived by bees (*right*). (**b**) How lights along the frequency spectrum are perceived by horses and by humans is shown (colors according to Carroll et al. 2001)

UV light[5] (Fig. 4.2a). Another example of animal species that have evolved the capacity to sense stimuli that humans are blind to is those species that can perceive electrical stimuli. Because water is a much better conductor of electricity than air, electroreception is most often found in aquatic animals. Sharks, for example, can sense the weak electric fields generated by the nerve and muscle activity of potential prey. Some shark species, such as the lemon shark, use this information to coordinate their attacks. Other species are active in the dark, where vision is not a useful modality to avoid bumping into things. Bats, for example, have solved this problem through echolocation, a sense that allows them to perceive the sound-reflecting (instead of light-reflecting) properties of their environment.

Just as some animal species have acquired the ability to sense properties of their environment that humans are blind to, other species never acquired some of the perceptual capacities of humans. Horses, for example, have a less-developed color discrimination ability than humans (Fig. 4.2b). Other species had certain perceptual capacities, but lost them again because they moved into an ecological niche in which there was no need to detect certain aspects of the physical world. Some species of cavefish, which live in dark caves, have lost their eyes and therefore their ability to perceive light (Jeffery 2009). Similarly, carnivorous mammals, which have a sugar-free, meat-only diet, do not need to perceive sweet tastants. Consequently, many species like the cat, the spotted hyena, and the fur seal, have lost their sweet taste receptor, Tas1r2 (Jiang et al. 2012).

One could argue that in all these cases the different species do not perceive the physical world differently; instead, they perceive different properties of the physical world. For example, bees, with their capacity to perceive UV light, perceive the UV-reflecting property of the physical world. The blind cavefish, in contrast, do not perceive any light-reflecting properties of the physical world. However, in many cases, different animal species perceive the same properties of the physical world, but they perceive these properties differently. Many examples have been discussed in Sect. 2.1, where I discussed the diversity of perceptual spaces. However, not only the perceptual qualities themselves can differ between different species. The spatial and temporal arrangements of the perceptual qualities can also differ. Visual perception in fruit flies, for example, has higher temporal but lower spatial resolution than human vision. The properties of the physical world that fruit flies and humans perceive visually overlap largely, but how the two species perceive the physical world visually differs dramatically.

Perceptual Variability Within the Same Species

How the physical world is perceived differs not only between different species. Even members of the same species can have different behavioral goals[6] and therefore different ways of perceiving their environment. In some species, for example, the perceptual systems differ between the sexes. The best-known examples are perceptual systems that are specialized for finding mates or facilitating courtship. Male moths, for example, have sensory organs that are specialized for the detection of the pheromones released by the females of the species. In other species, perceptual systems that are not involved in guiding sexual behaviors are sexually dimorphic. Some species of anglerfish, for example, show extreme sexual dimorphism (Pietsch 2009). The males of these species are much smaller than the females and spend most of their life as parasitic appendages attached to a much larger female in permanent parasitic conjugation. The males have very different behaviors from that of the females and therefore perceive the physical world very differently.

Anglerfish also illustrate that even within the same individual, behaviors and therefore the demands on perception can change. Before the males attach to the female, they have large eyes that are specialized for detecting the bioluminescent lure of the female. After they have used these large eyes to find a female to attach to, they no longer need to move independently. Their eyes, like the eyes of the females, degenerate.

Changes in an individual's perception are also common in less exotic animals. Consider, for example, the fruit fly, which lives as a larva before it pupates and then emerges as an adult fly. The behaviors that perception needs to guide for larvae are very different from the behaviors that perception has to guide in adults. The larva hatches from the egg on rotten fruit and spends the entire time before pupation eating. The adult fly that emerges from the pupae has a much richer repertoire of behaviors that includes finding a mate, mating, finding an appropriate site to lay eggs, laying eggs, and so on. Consequently, the same individual has a much richer perception of its environment as an adult. Most notably, adults fly whereas larvae cannot fly. The high temporal resolution of visual perception that flies need to execute flight maneuvers is therefore only found in the adult.

In other cases, perception changes with the behavioral demands within the same individual in the same developmental stage. Female mosquitoes, for example, have to find an animal to bite because blood is their main source of protein. They need protein to produce eggs. After filling up with blood, the mosquito has to find a puddle of water to lay eggs in. This abrupt change from having to find a source of blood to having to find a place to lay eggs is mediated by a change in the olfactory sensory neurons of the animal. These neurons become less sensitive to body odors and more sensitive to odors given off by potential egg-laying sites (Davis 1984). Similar changes in the activity of the most peripheral olfactory sensory neurons have also been found in other species (see, e.g., Root et al. 2011; Saveer et al. 2012).

In humans, a good example to illustrate the dependence of perception on changes in behavioral goals is the change in olfactory perception during pregnancy (for a review, see Doty and Cameron 2009). A large number of self-reports indicate that the perception of food smells changes during pregnancy. The reason for this is that the most adaptive eating behavior is different during pregnancy. A pregnant woman has to not only feed herself, but also the embryo growing in her. An adult usually eats for energy, but during pregnancy, there is rapid tissue growth that needs to be sustained with a different combination of nutrients. In addition to the need to fuel the embryonic growth, pregnant women also need to be more vigilant than usual to avoid food that, although save for adults, might endanger the embryo. The most adaptive behaviors toward the same food items are therefore different during pregnancy as during other times. Consequently, women perceive the smell of these food items differently during pregnancy.

Different perceivers all encounter the same physical world. How they perceive this world differs dramatically. The largest differences are found between members of different species, but members of the same species can also differ substantially in how they perceive the same stimulus. If correctness of perception is based on a relation between what is perceived and how it is perceived, then there are instances in which animals evolve to perceive the physical world less accurately, which is a strong indication that perceiving the world accurately is not the evolutionary function of perception.

4.2 Similarity of Percepts and Similarity of Stimuli

The philosopher Thomas Nagel writes in his discussion of the function of perception:

> Perception and desire have to meet certain standards of accuracy to enable creatures to survive in the world: they have to enable us to respond similarly to things that are similar and differently to things that are different, to avoid what is harmful, and to pursue what is beneficial (Nagel 2012, p. 73).

Nagel here mentions two standards that perception has to meet. First, perception has to enable us to respond similarly to similar things. Second, perception has to enable us to avoid what is harmful and pursue what is beneficial. These two standards are alternative formulations of the two proposals about the function of perception that are compared in this chapter. Retaining similarity relations in the physical world in perception is the criterion for accurate or correct perception. Avoiding harmful things and pursuing beneficial things is the criterion for guiding adaptive behaviors. Plainly, the two standards mentioned by Nagel are often in conflict with one another. The deadly coral snake and the harmless milk snake are very similar in their visual appearance. However, one is harmful and the other, for a hungry hunter, beneficial. We cannot simultaneously respond to these two visually similar snakes similarly, while also avoiding what is harmful and pursuing what is beneficial.

Situations in which perceiving what is physically similar as similar is in conflict with perceiving things that are harmful as different from things that are beneficial are interesting for understanding what the function of perception is. Examining such situations will reveal whether perception prioritizes guiding behaviors or collecting correct information. When evolutionary processes can shape our perceptual systems either to ensure that we avoid what is harmful and pursue what is beneficial *or* to respond similar to things that are similar, what is the outcome? In this section, I will show that the outcome is that evolution through natural selection selects the perceptual system that ensures adaptive behaviors over the alternative perceptual system that ensures correctness of the collected information. The chemical senses, especially the perception of bitter tastants, provide good case studies to illustrate this outcome.[7]

Bitter Tastants

Smell, taste, and other modalities combine to form the experience of flavor. One of the basic tastes that contribute to flavor is bitter taste. Extremely diverse chemicals, like hydrolyzed proteins, alkaloids, rancid fats, and poisons (Martin 2013, p. 65), all have a bitter taste (Fig. 4.3). These bitter tastants are produced by a large variety of different plants. What most of the

sucrose octaacetate denatonium benzoate quinine

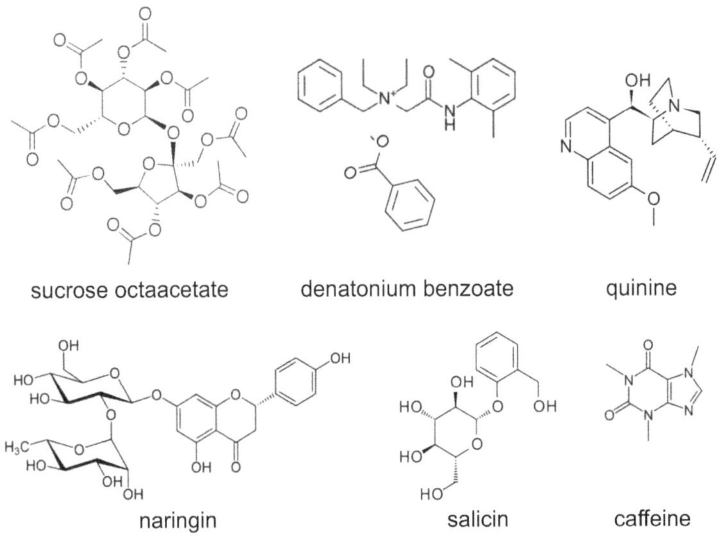

naringin salicin caffeine

Fig. 4.3 Six structurally diverse molecules that are perceived as bitter (Laska et al. 2009)

plants in which bitter tastants are found have in common is that they are inedible or even toxic. Consequently, whenever you perceive a bitter tastant, the appropriate behavioral response is to spit out what is in your mouth. Bitter tastants are therefore an excellent example of physically diverse stimuli that all require the same behavioral response. The perceptual qualities associated with these stimuli correlate much stronger with the appropriate behavioral response than with any physical feature of the stimuli. Hydrolyzed proteins and rancid fats are perceived as being similar, although they are physically very different. Non-rancid fat is physically more similar to rancid fat than hydrolyzed proteins. However, because rancid fat and hydrolyzed proteins have reduced nutritional value and because rancidification can produce toxic compounds, they are perceived as bitter. Because non-rancid fat is an excellent source of energy, it is not perceived as bitter. Perception does not reflect the physical similarity between the stimuli. This shows that, when efficiently guiding behaviors involves misrepresenting similarity relations between stimuli, as it is the case in bitter perception, then perception will misrepresent the similarity relations found in the physical world.

The correlation between bitter perceptual qualities and the appropriate behavioral response of spitting out whatever is perceived is not perfect. Sometimes, bitter foods and drinks are incorporated into our diet, which seems to contradict the idea that bitter taste is a signature of inedibility. Theobromine in chocolate and caffeine in coffee are two examples of bitter alkaloids consumed by humans. However, these chemicals are not examples of non-toxic bitter tastants. Instead, we incorporated them into our diet *because* they are toxic. At low enough doses, these toxins, like many others, act as stimulants. To bypass the evolved response to bitter substances, large amounts of sugar can be added to coffee or chocolate to make bitter stimulants palatable.

Other bitter compounds are edible even at high doses. There is an ongoing arms race between those who try to eat plants, and plants that try to avoid being eaten. A good defense against being eaten is to become toxic. Once toxic plants evolved, an equally good defense against being eaten is to taste like a toxic plant. This strategy of imitating the warning signal of a harmful species is known as Batesian mimicry. The similarity between harmless milk snakes and venomous coral snakes is the result of Batesian mimicry. Bitter tasting compounds that are edible may also be the result of Batesian mimicry. To avoid being eaten by humans, announcing that one is inedible through bitter taste is sufficient as long as many plants that taste bitter are actually inedible. Because our perceptual systems coevolve with the natural environment they perceive, the correlation between the perception of bitterness and the optimal behavioral response is not perfect. Some inedible compounds are not bitter and some bitter compounds are edible. However, the correlation between the perceptual quality and the appropriate behavior is better than the correlation between the perceptual quality and the physical properties of the stimulus.

Evolution of Perceived Similarity

One can speculate how chemical sensors that are sensitive to a wide variety of different chemicals that are physically different but require the same behavioral response, like the bitter tastant receptors, have evolved. The first step in the perception of tastants is that they are bound by a receptor molecule on the surface of a sensory neuron. There are different types of receptor molecules that bind to different types of tastants. Whether a given type of receptor molecule binds to a tastant or not depends on the receptor's molecular structure. The molecular structure of the receptor is determined by its amino acid sequence, which in turn is determined by the DNA sequence of its gene. Mutations in the receptor gene can alter the structure of the receptor and therefore change which chemicals can activate the receptor. Imagine that the first receptor in evolutionary history that bound a bitter tastant did bind to solanine, a bitter tasting toxin found in many plants. To be adaptive, activation of the receptor must have triggered the spitting out of the solanine-containing food that activated the receptor. Over time, random mutations have produced slight variants of this receptor. If any of these variants gained sensitivity to a second chemical that is found in a different inedible plant species, then the mutation was advantageous, because it increased the ratio of edible over inedible plants consumed. If the mutation made the receptor sensitive to a chemical found in edible plant species, it resulted in the edible plant being not consumed. This mutation therefore decreased the ratio of edible over inedible plants consumed and therefore was non-advantageous. In this manner, a receptor that detects the difference between edible and inedible plants can evolve. "Edible" and "inedible" are not features of the physical world, but they are properties that are relative to the perceiver.[8] As the hypothetical bitter receptor evolved into a reliable indicator of whether the perceiving organism should swallow or spit out the perceived food, it evolved away from accurately representing similarity of chemicals. The stimuli the receptor responded to became more and more physically diverse.

The mechanistic explanation of the evolution of a receptor sensitive to diverse molecules that all require the same behavioral response illustrates how a chemosensory system that guides behavior without collecting accurate information about similarity relations in the physical world can evolve from a system that collects accurate information. The mechanistic explanation also illuminates why cases like this are more likely to be found in the chemical senses than in vision. The reason for this difference is that the visual system perceives identity and location of objects, whereas the olfactory system only perceives identity of odorants. Just as many taste molecules require the same behavioral responses, there are instances of visual perception in which different visual objects require the same behavioral response. Whether somebody throws a rock, a book, or a shoe at you, the right response in all three cases is to move out of the object's trajectory. The same behavioral response is required regardless of *what* is perceived, just like in the case of bitter tastants. However, the required behavioral response to having any object thrown at you depends strongly on *where* the object is perceived. If it is perceived at eyelevel in the front, ducking is the best response. If it is perceived lower, stepping to the side will be more appropriate. If it is perceived to the left, the left arm should be raised to cover the face. If it is perceived to the right, the right arm should be raised to cover the face. When it comes to the perception of location, each location requires a different behavioral response because behaviors that are responses to the location of a perceived object are directed behaviors.

Consider a fictional situation to illustrate how the need for accurate presentation of the location of objects in vision results in a perceptual system in which similarity in perception often reflects similarity in the physical world. Let us assume that the presence of a lion, a rattlesnake, and an alligator all require the same behavioral response: freezing. Being able to discriminate these three animals is not advantageous because the optimal behavior in response to detecting them is the same. Now, if the three animals are perceived olfactorily, the olfactory system may evolve into a system that does not discriminate between lions, snakes, and alligators. An odorant receptor that mediates the freezing behavior could evolve to bind the different odors given off by the three different animals. We would lose the ability to discriminate these animals olfactorily because discriminating them has no adaptive value.

Now consider an alternative scenario. Assume that the appropriate behavioral response to lions, rattlesnakes, and alligators is running away from them. Further, assume that the animals are perceived visually. In this situation, we would not only need to know that we are encountering a lion, rattlesnake, or lion, but also where the animal is. Running *away* from it requires locating it in space. In this situation, the visual system would not lose the ability to discriminate the three animals, although discriminating them has no adaptive advantage. The reason for this is that how the animal's identity is perceived is severely constrained by the need to perceive its location. A system that has to perceive the location of objects will automatically also perceive the shapes and the movements of objects, and rattlesnakes are shaped very differently from lions. They therefore also move differently through the visual space. This is why in vision and other modalities in which spatial perception plays an important role, guiding behaviors adaptively and collecting accurate information about the physical world often coincide.

The perception of space (and the perception of time, which I have not discussed here) is a notable exception to the rule that the similarity of percepts does not reflect similarity in the perceived stimuli but similarity in the adaptive behavior that the stimuli require. The reason for this exception is that successfully executing behaviors directed toward objects in the physical space requires accurate information about the position of the objects in physical space.

4.3 Alternative Notions of Correctness

Members of different animal species perceive the same physical stimuli vastly differently. Even individuals of the same species evolved to perceive the same stimulus differently depending on their behavioral goals. Furthermore, when perceiving similar things similarly is in conflict with perceiving harmful things to be different from beneficial things, perceptual similarity reflects behavioral relevance rather than physical similarity of the stimuli. This type of evidence shows that perception is not about perceiving the physical world correctly, but about guiding beneficial behaviors. However, this conclusion only holds when *correct perception* is defined as perception in which the physical similarities in the world are reflected in similarities between percepts. A possible way to continue to resist a behavior-based function of perception in light of this evidence is therefore to adjust the notion of perceptual correctness. Two alternative notions of correctness have been suggested in this context.[9]

Perceptual System-Dependent Correctness

One alternative way of thinking about correctness in the context of perception is to define correct perception as perception that is triggered by its ordinary causes. When this definition is applied, then the gustatorily perception of bitterness is correct if it is triggered by the molecules that ordinarily activate the bitter receptor. This is akin to the proposal that perception represents a disjunctive property that is probably open-ended (Lycan 2014). The information that is collected by the perception of bitterness therefore is "What I currently have in my mouth is either sucrose octaacetate, or denatonium benzoate, or quinine, or naringin, or salicin, or caffeine, or ...". This information is correct when I have one of these chemicals in my mouth. An alternative and shorter way of saying the same thing is that the information about the physical world that I collect when perceiving bitterness is "What I currently have in my mouth is capable of activating my bitter receptor". This information is correct when something I have in my mouth is capable of activating the bitter receptor. Correctness of perception, under this notion, is dependent on the perceptual system. As the human bitter receptor evolves, the molecules it binds change. However, under the notion of perceptual system-dependent correctness, this change does not result in a change in how correct the receptor perceives the physical world. The old version of the receptor and the new version of the receptor perceive the world differently, but both perceive it correctly. The visual perception of an animal with a single photosensitive cell is as correct as the visual perception of the animal with the most sophisticated apparatus. Indeed, all perception by a properly functioning perceptual system is by definition correct perception when correctness is defined in terms of the perceptual system (Akins 1996).

Perceiver-Dependent Correctness

A notion of perceptual correctness that, like perceptual system-dependent correctness, does not rely on the relation between the content of the perception and the stimulus is perceiver-dependent correctness. The difference between perceptual system-dependent correctness and perceiver-dependent correctness is that perceptual system-dependent correctness depends on the interaction between the stimulus and the perceptual system, whereas perceiver-dependent correctness depends on the interaction between the stimulus and the perceiver. The perception of bitterness is perceptual system-dependent correct when the perceived stimulus is capable of activating the bitter receptor. It is perceiver-dependent correct when the stimulus is inedible or toxic for the perceiver. In the first case the correctness depends on facts about the perceiver's perceptual system; in the second case, it depends on facts about the perceiver. This difference can be illustrated using the example of artificial sweeteners. The perceptual quality of sweetness is usually associated with tastants that have high caloric value. The appropriate behavioral response to it is to swallow. Sugars like glucose are typical examples of chemicals that taste sweet and have high caloric value. Artificial sweeteners are molecules that taste sweet but have low caloric value. Under the notion of perceptual system-dependent correctness, perceiving artificial sweeteners as sweet is correct perception because the perception is triggered by molecules that ordinarily activate the sweet receptor. Under the notion of perceiver-dependent correctness, the perception of artificial sweeteners as sweet is not correct because the behavior triggered by sweet perception (swallowing) is not adaptive for molecules with no caloric content. The reason why the sweet receptor misinforms us about the caloric content of artificial sweeteners is that these non-natural molecules were not present in the environment in which the perceptual system evolved.

Matthen has suggested a version of perceiver dependence of perceptual correctness in the form of species-specific standards of correctness (Matthen 2005). According to Matthen, every species has different perceptual systems and different ecological needs and therefore needs to perceive the physical world in a way specific to that species. As was reviewed above, perceptual systems and perception not only differ between different species but also between the sexes and life stages of the same species. Species-specific standards of correctness are therefore not enough. What is required are species-, sex-, life stage-, situation-specific standards of correctness.

Perceptual system dependence and perceiver dependence of perceptual correctness both propose that correctness of perception is independent of the relation between the stimulus and the percepts. Instead of a mechanism that collects correct information about the stimulus, perception becomes a mechanism to collect correct information about the relation between the perceiver and the physical world. The molecule binds to *my* bitter receptor. The bitter plant is inedible *for me*.

4.4 Conclusion: Perception Evolved for Guiding Behaviors

I have set this chapter up as a comparison between two proposals about the evolutionary function of perception: collecting correct information about the environment and guiding behaviors. Investigating chemosensory perception reveals that perception did not evolve to correctly represent, map, or collect information about the physical world. Instead, it evolved to guide adaptive behaviors of the perceiver. Adaptive behaviors are driving evolution with no regard to the correctness of the underlying percepts, beliefs, or calculations.

The impression that perception guides behavior *by* collecting accurate information about the physical world is most seductive when considering the visual perception of space. Successfully hunting a boar requires accurate information about the position of the boar in space so that boar-directed movements can be executed. Other aspects of visual perception are constrained by this need for accurate spatial representation, and much of visual perception therefore accurately reflects similarities in position and shape. As I have shown, collecting accurate information about the physical world is not a function of chemosensory perception or other forms of non-spatial perception. The only function of perceiving smells and tastes is guiding adaptive behaviors. Collecting accurate information by itself has no adaptive value, and when perception guides adaptive behaviors by other means, collecting correct information provides no additional benefit for the organism.

The conclusion that collecting correct information is not the function of perception depends on a notion of correctness that defines correct perception as perception in which similarities between stimuli are reflected in similarities between percepts. Some scholars who resist the behavior-based function of perception have defended the collection of correct information as a function by redefining perceptual correctness. According to these alternative notions of perceptual correctness, whether a given instance of perception is an instance of correct perception does not depend on facts about the stimulus, but on facts about the perceiver. For example, different species need to respond differently to different stimuli. Whether a given chemical is toxic for the perceiver depends on the perceiver's physiology. A substance that is toxic for a human is not necessarily toxic for a trout and the other way around. The similarity in perception does not reflect the similarity of the chemicals, which would be the same for all perceivers, but the similarity in toxicity, which depends on the perceiver and differs between different species. Ultimately, these perceiver-dependent accounts of correctness define correctness as being dependent on behavior: the correct perception of something toxic is the perception that results in the evolved adaptive response to toxins.

Notes

1. There are diverse philosophical theories about what "functions" in biology are (for a collection of essays on the topic, see Buller (1999). *Function, Selection, and Design*. Albany, SUNY Press). Functions can be either teleological functions, or causal (or systemic) functions. The teleological function of a biological structure or mechanism is what it was selected for. Millikan (1984). *Language, Thought, and Other Biological Categories*. Cambridge, MIT Press. The causal function of a mechanism is the role of a structure or mechanism within a complex system. Cummins (1975). "Functional analysis." *The Journal of Philosophy* **72**: 741–765. The discussion here is about the teleological function of perception. The teleological function is supposed to explain why something is there. The teleological function of something is the same as its adaptive value or the reason why it was selected.

2. I use "accurate" and "correct" as synonyms. What they mean in the context of perception will be discussed in section 4.3. I chose "collecting accurate information" as the alternative possibility of what perception's non-behavioral function cold be. However, there are many similar proposals. One suggestion would be that perception *maps* the physical world accurately, another that it *represents* the physical world accurately. My argument that it is the function of perception to guide behaviors works against all of these non-behavioral proposals equally.
3. Evolutionary biologists call by-products of the evolution of other characteristics that themselves are not direct products of adaptive selection "spandrels". Gould and Lewontin (1979). "The spandrels of San Marco and the Panglossian Paradigm: A critique of the adaptationist programme." *Proceedings of the Royal Society of London, Series B* **205**(1161): 581–598.
4. Helping riders navigate the subway system is the function of the subway maps on display in subway stations. One can imagine other subway maps with other functions. There could be a map of the subway system used by the fire department for emergencies. In this map, geographical accuracy is important and it is irrelevant how helpful the map is for tourists navigating the subway system.
5. The flowers profit from providing nectar for bees because nectar-collecting bees carry pollen between flowers, thereby pollinating the flowers. The bees' capacity to detect the UV patterns of flowers and the UV patterns of flowers presumably coevolved.
6. Most commonly, the differences in goals are between males and females, but they also can be between parents and offspring.
7. The perception of temperature also provides excellent examples that have been discussed in detail by Akins. Akins (1996). "Of sensory systems and the "aboutness" of mental states." *The Journal of Philosophy* **93**(7): 337–372. As Akins shows, the thermoreceptive system does not have the function of measuring the temperature in the environment as accurately as possible. Instead, its function is to help the perceiver execute behaviors that aim at avoiding damage through cold or heat. Evidence that this is the case is that the same thermal stimulus is perceived differently with different parts of the skin. Areas of skin that cover structures that are more delicate "exaggerate" temperature extremes. Similarly, the endpoint of (more dangerous) rapid temperature changes are perceived to be of different temperature from when the same endpoint has been reached through (less dangerous) gradual temperature changes.

8. Whether a compound is edible and how nutritious it is for a given organism depend on the organism's digestive system and physiology. Cows have a specialized stomach for digesting grass. Termites can digest wood with the help of symbiotic bacteria in their guts. Neither grass nor wood are good food sources for humans who lack the adaptations to make use of these food types.
9. Both of these possible responses are discussed by Akins as potential defenses of what she calls "the traditional view". She calls the first possibility I discuss in this section "the *a priori* defense" and the second the "appeal to biologically salient properties" Akins (1996). "Of sensory systems and the "aboutness" of mental states." *The Journal of Philosophy* **93**(7): 337–372.

References

Akins, K. (1996). Of sensory systems and the "aboutness" of mental states. *The Journal of Philosophy, 93*(7), 337–372.

Buller, D. J. (Ed.). (1999). *Function, selection, and design*. Albany: SUNY Press.

Carroll, J., Murphy, C. J., et al. (2001). Photopigment basis for dichromatic color vision in the horse. *Journal of Vision, 1*(2), 80–87.

Cummins, R. (1975). Functional analysis. *The Journal of Philosophy, 72*, 741–765.

Davis, E. E. (1984). Regulation of sensitivity in the peripheral chemoreceptor systems for host-seeking behavior by a hemolymph-borne factor in Aedes aegypti. *Journal of Insect Physiology, 30*, 179–183.

Doty, R. L., & Cameron, E. L. (2009). Sex differences and reproductive hormone influences on human odor perception. *Physiology & Behavior, 97*(2), 213–228.

Gould, S. J., & Lewontin, R. C. (1979). The spandrels of San Marco and the Panglossian Paradigm: A critique of the adaptationist programme. *Proceedings of the Royal Society of London Series B, 205*(1161), 581–598.

Jabbour, E. (2010). Mapping information: Redesigning the New York City subway map. In J. Steele & N. Iliinsky (Eds.), *Beautiful visualization: Looking at data through the eyes of experts* (pp. 69–89). Sebastopol: O'Reilly Media.

Jeffery, W. R. (2009). Regressive evolution in Astyanax cavefish. *Annual Review of Genetics, 43*(1), 25–47.

Jiang, P. H., Josue, J., et al. (2012). Major taste loss in carnivorous mammals. *Proceedings of the National Academy of Sciences, 109*(13), 4956–4961.

Laska, M., Rivas Bautista, R. M., et al. (2009). Gustatory responsiveness to six bitter tastants in three species of nonhuman primates. *Journal of Chemical Ecology, 35*(5), 560–571.

Lycan, W. G. (2014). The intentionality of smell. *Frontiers in Psychology, 5*, 436.

Martin, G. N. (2013). *The neuropsychology of smell and taste*. London: Psychology Press.

Matthen, M. (2005). *Seeing, doing, and knowing*. Oxford: Oxford University Press.

Millikan, R. (1984). *Language, thought, and other biological categories*. Cambridge: MIT Press.

Nagel, T. (2012). *Mind & Cosmos: Why the materialist Neo-Darwinian conception of nature is almost certainly false*. Oxford: Oxford University Press.

Pietsch, T. W. (2009). *Oceanic Anglerfish: Extraordinary diversity in the deep sea*. Berkeley: University of California Press.

Root, C. M., Ko, K. I., et al. (2011). Presynaptic facilitation by neuropeptide signaling mediates odor-driven food search. *Cell, 145*(1), 133–144.

Saveer, A. M., Kromann, S. H., et al. (2012). Floral to green: Mating switches moth olfactory coding and preference. *Proceedings of the Royal Society B: Biological Sciences, 279*(1737), 2314–2322.

Part 3

Olfaction and Cognitive Processes

So far, I have discussed the perception of perceptual qualities. I have also explored how these qualities can be combined in time and space, whether these combinations amount to something like perceptual objects, and what the function of perceiving and combining perceptual qualities is. Throughout, I have pretended that perception is separate from other cognitive processes and that the different perceptual modalities are not connected. Now I will acknowledge that the distinction between perceptual and non-perceptual processes is largely arbitrary. Therefore, a satisfying account of olfactory perception has to include an analysis of the relation between olfaction and non-perceptual cognitive processes, as well as of the relation between olfaction and other perceptual modalities.

In Chap. 5, I will discuss the flow of information from the olfactory system to cognitive processes. I will argue that emotional processes have privileged access to olfactory information whereas the connection between olfaction and language is weak. I will contrast this finding with the availability of visual information throughout the mind and argue that the connection between olfaction and emotional processes reflects the evolutionary history of olfactory perception. Because each modality has a different evolutionary history, it is not possible to generalize from the connectivity observed in one modality to the connectivity of all other modalities.

In Chap. 6, I will then address the flow of information from other parts of the mind towards the olfactory system. I will first discuss the flow of information from cognitive processes towards olfaction, a process often called cognitive penetration of perception. In the second part of Chap. 6, I will address input into the olfactory system from other perceptual systems.

Throughout this discussion of the structure of the mind, I will consider the mind (much like the brain) as a collection of overlapping networks rather than as an assembly of modules and a non-modular remainder.[1] The task of discovering the structure of the mind is the task of discovering the connections that constitute these networks. One possibility is to use introspection. When smelling food is more likely to make us hungry than touching food, this can be interpreted as evidence for a closer connection between olfaction and hunger regulation than between touch and hunger regulation. However, introspection has two disadvantages. One disadvantage is that a project to elucidate connections in the mind based on introspection would come up with a very impoverished structure. Just because we are not aware that one process influences another process does not mean that there is no such influence. Using introspection will therefore miss many of the connections in the mind. Secondly, there is no procedure for weighing the evidence in cases of disagreeing introspections between individuals.

Another method to discover networks in the mind is behavioral experiments. Behavioral experiments avoid to some degree the two problems of introspective evidence. When touching food makes subjects eat more food in a behavioral experiment, then this is evidence for a connection between the parts of the mind that process tactile information and the parts of the mind that regulate food intake, regardless of the subjects' self-reports about the existence of such a connection. Furthermore, in behavioral experiments the sample size is larger than in introspection and it is therefore possible to quantify the extent of inter-individual variability. Connections in the mind are likely to be partly hardwired and partly shaped by previous experiences and it is interesting to distinguish innate and acquired structure.

In addition to introspection and behavioral experiments, neuroanatomy, and especially functional neuroanatomy, can also provide evidence about the structure of the mind. Like the other sources of evidence, functional neuroanatomy has its disadvantages. The most troubling problem with drawing conclusions about the structure of the mind from observing the brain is that it presupposes a close association between brain areas and parts of the mind. That there is such an association is obvious, especially in sensory areas. Activity in the olfactory bulb is much more likely to be associated with smelling than with seeing whereas the visual cortex is most active during visual perception. Many structure-function associations have also become apparent in the central brain. Activity in the amygdala is more likely to be associated with experiencing fear than with counting. However, the brain regions corresponding to some parts of the mind may be spatially distributed rather than concentrated in one region of the brain, or the regions may be too small to be resolvable with current methods.

Note

1. In cognitive science, "module" is a technical term that is defined differently by different authors. The terminology of modularity of the mind goes back to Fodor. Fodor, J. (1983). *The Modularity of Mind.* Cambridge, MIT Press. Fodor defined modules as those parts of the mind that satisfy nine criteria. Of the original nine criteria, the two criteria that continue to be evoked most frequently are informational encapsulation and domain specificity. Samuels, R. (2006). Is the human mind massively modular? *Contemporary Debates in Cognitive Science.* R. Stainton, Blackwell. A part of the mind is informationally encapsulated if it does not receive information from higher centers of processing in the brain. A part of the mind is domain specific if it processes a restricted type of information, for example, olfactory information or information about linguistic rules. Cosmides, L. and J. Tooby (1994). Origins of domain specificity: The evolution of functional organization. *Mapping the Mind: Domain Specificity in Cognition and Culture.* L. A. Hirschfeld and S. A. Gelman. New York, Cambridge University Press: 85–116, Sperber, D. (1994). The modularity of thought and the epidemiology of representations. *Mapping the mind: Domain*

Specificity in Cognition and Culture. L. A. Hirschfeld and S. A. Gelman. New York, Cambridge University Press.: 29–67. For a more detailed discussion of modules, see Robbins, P. (2010). "Modularity of Mind." *The Stanford Encyclopedia of Philosophy* Summer 2010. from http://plato.stanford.edu/archives/sum2010/entries/modularity-mind/, and for a critique of the concept see Prinz, J. J. (2006). Is the mind really modular? *Contemporary Debates in Cognitive Science*. R. Stainton: 22–36. How the modular structure of the mind is described depends on the definition of "module" that is employed and I am not aware of a principled way to decide whether the modular structure based on informational encapsulation reflects the actual structure of the mind better than the modular structure based on domain specificity. I therefore support Jesse Prinz's suggestion that, although it is uncontroversial that there are systems in the mind that carry out distinct functions, the term "modularity" should be dropped (ibid.).

References

Cosmides, L., & Tooby, J. (1994). Origins of domain specificity: The evolution of functional organization. In L. A. Hirschfeld & S. A. Gelman (Eds.), *Mapping the mind: Domain specificity in cognition and culture* (pp. 85–116). New York: Cambridge University Press.

Fodor, J. (1983). The modularity of mind. Cambridge: MIT Press.

Prinz, J. J. (2006). Is the mind really modular? In R. Stainton (Ed.), *Contemporary debates in cognitive science* (pp. 22–36). Oxford: Blackwell.

Robbins, P. (2010). Modularity of mind. *The Stanford Encyclopedia of Philosophy* Summer 2010. From http://plato.stanford.edu/archives/sum2010/entries/modularity-mind/

Samuels, R. (2006). Is the human mind massively modular? In R. Stainton (Ed.), *Contemporary debates in cognitive science*. Oxford: Blackwell.

Sperber, D. (1994). The modularity of thought and the epidemiology of representations. In L. A. Hirschfeld & S. A. Gelman (Eds.), *Mapping the mind: Domain specificity in cognition and culture* (pp. 29–67). New York: Cambridge University Press.

5

Availability of Olfactory Information for Cognitive Processes

Olfaction is often considered the most animalistic and primitive of our senses. Odor stimuli induce desires, emotions, and physiological responses that make us respond to certain smells in automatic ways. Reason is powerless to intervene. In contrast, it is difficult to talk about smells, or even to name them. In this chapter, I will show that these peculiarities of olfaction are based on differences in how well connected olfaction is to cognitive processes involved in evaluative emotions and in language, respectively. The results of behavioral experiments, as well as neuroanatomical and functional evidence, demonstrate that olfaction has a privileged connection to evaluative emotional processing. On the other hand, the information flow from olfaction to the language centers is comparably weak.

5.1 Olfaction and Language

> The sight in my opinion is the source of the greatest benefit to us, for had we never seen the stars, and the sun, and the heaven, none of the words which we have spoken about the universe would ever have been uttered.

A. Keller, *Philosophy of Olfactory Perception*,
DOI 10.1007/978-3-319-33645-9_5

> But now the sight of day and night, and the months and the revolutions of the years, have created number, and have given us a conception of time, and the power of enquiring about the nature of the universe; and from this source we have derived philosophy, than which no greater good ever was or will be given by the gods to mortal man. This is the greatest boon of sight. Plato's *Timaeus*

Plato tells us that visual perception is a requirement for language; without sight, "none of the words we have spoken about the universe would ever have been uttered".[1] Language, in turn, is the tool of philosophy. Without vision, there would therefore be no philosophy, which is probably why philosophers concerned with perception have such a strong preference for vision over other modalities. Modern psychology and neuroscience have confirmed that Plato's intuition has some truth to it. It is easier for us to name and talk about colors than to name and talk about smells. There are two aspects of this difficulty to talk about smells. One problem is the lack of a smell vocabulary. Many languages have words for colors, like "blue" and "green" (Berlin and Kay 1969). At least the English language does not have equivalent words for smells. Words used to describe smells are either judgments about the smell and its effects ("horrid", "soothing"), or, most frequently, the name of the source ("flowery", "leathery"). Why do we lack a smell vocabulary? It is possible that the lack of a smell vocabulary is caused by cognitive architecture. However, it is also possible that language coding is, for some reason, better suited to express some sensations rather than others. A third alternative explanation is that the cultural forces that shaped language happened to shape English in a way that reflects the relatively higher importance of colors compared to smells for the culture in which it evolved (for a review of these three possible explanations and of modality-dependent ineffability in general, see Levinson and Majid 2014). Because there are alternative explanations, the difference between our smell vocabulary and our color vocabulary does not show that there is an impoverished connection between olfaction and language centers. However, our limited abilities to talk about smells are not only due to the lack of an appropriate vocabulary. A second problem with talking about smells is that, even when there is an appropriate word to label a smell, we often fail to access it. This inability to access language to name smells or talk about them provides the evidence for a poor connection between olfactory processing and the language center that I will discuss in this section.

Naming Smells

It is difficult for us to name a smell. To some degree, this is because, during development, we form much fewer associations between smells and verbal labels than between sights and verbal labels. Adults spend considerable time with preverbal children looking at picture books and pointing at drawn objects while saying, "this is a fire truck" or "this is a cow". Much less time is spent holding odors under children's noses while uttering the odors' names. Consequently, most of us have many more associations between visual appearances and names than between smells and names. However, for some odors, like coffee, sweat, gasoline, or garlic, there have been (for people with life histories similar to mine) many chances to learn the name of the odor. Interestingly, even for those very common and familiar odors, naming the odor is astoundingly difficult. In one experiment, the majority of participants were unable to name the smells of beer, urine, roses, or motor oil (Desor and Beauchamp 1974).

The inability to name an odor can have different reasons. The process of naming odors, just like any naming process, consists of three steps. First, the odor has to be identified. After the odor has been identified, the verbal label that is associated with the odor has to be activated. Finally, the response has to be generated (Johnson et al. 1996). The identification step can be further subdivided. To identify an odor, it has to be detected, discriminated from other odors, and recognized. Odor recognition consists in matching the perceived smell to a previously perceived smell. Recognition does not imply the ability to name. One can recognize an actress in a movie from having seen her previously in another movie without being able to name her (Chobor 1992, p. 356). The poor performance of subjects in odor-naming experiments could be due to difficulties at any of the steps involved in the naming process. The most likely explanation for the difficulties with odor naming is that accessing linguistic semantic information about odors is difficult (for the evidence that this is the case, see Stevenson 2009). This would mean that the subjects that show poor odor-naming abilities in experiments are able to recognize the odors, but are unable to name them. To test whether this is true, one would have to perform a test of odor recognition that does not depend on verbal report. For example, one could ask subjects who cannot name the odors of motor oil, urine, or beer, which one of the three they would rather drink. My prediction is that subjects would decide to drink beer more frequently than urine or motor oil. Similarly, I predict that they would be unlikely to pour the beer in their car's engine. If these predictions are true, then the deficits in odor naming are due to the difficulty of accessing linguistic labels for the odors. Either way, the failure to name an odor cannot reveal whether the odor has been identified correctly or not. Naming requires that, in addition to identification of the smell, the associated verbal label is activated and the response generated.

That the poor performance in odor naming is not due to problems in identifying the odor, but due to problems in making the connection between the perception and the appropriate verbal label is illustrated by the prevalence of the tip-of-the-nose phenomenon (Sulmont-Rosse 2005). The tip-of-the-nose phenomenon occurs when people are incapable of retrieving from memory the word that is associated with an odor, although they correctly identified the odor. The tip-of-the-nose phenomenon is named in analogy to the tip-of-the-tongue phenomenon, which is the failure to retrieve a word from memory in combination with the feeling that retrieval is immanent (Schwartz and Metcalfe 2011). Tip-of-the-nose phenomena are not caused by problems with odor identification, but by our inability to name odors. This is demonstrated by experiments in which subjects fail to name an odor correctly, but after they are provided with a list of odor names that includes the name of the odor that they have to name, or with other semantic information about the odor, they can name the odor (Sulmont-Rosse 2005; Gilbert 2008, p. 127).

Talking About Smells

It is very difficult to name an odor, even for somebody who knows the odor's name and does recognize the odor. Another striking difference between olfaction and vision with respect to language is how difficult it is to say anything about an odor that we recognize but cannot name. In vision, we commonly talk about things we cannot name. We can talk about someone's visual appearance and behavior without knowing his or her name. In fact, knowledge of the person's name would not make a difference in what we are able to say about them. In vision, when an object cannot be named, it is still possible to retrieve a large amount of information about the object from memory (Lambon Ralph et al. 2000). We can describe the appearance of an actor whose name is on the tip of our tongue. We can list the movies he was in and describe his appearance in the hope that somebody else will help us out and provide the name of the actor that we currently cannot access. In olfaction, this is not the case. Very little can be said about an odor unless we are able to name it (Jönsson et al. 2005). Stevenson writes: "What this suggests is that access to semantic information in vision is partially (if not fully) independent of the ability to name an object, while for olfaction a name appears necessary to access the same store of semantic information" (Stevenson 2009, p. 1008). It can be argued whether Stevenson is right and the problem is access to semantic information or more specifically access to linguistic semantic information. In an experiment that compared perfume experts with novices, it has been shown that the ability to perform actions that depend on semantic information like grouping of perfumes is to some degree independent of the ability to apply linguistic descriptors to those same perfumes (Veramendi et al. 2013). This, like the speculation above that even subjects who are not able to name the odors of beer and motor oil are unlikely to drink motor oil instead of beer, suggests that it is not all semantic information, but specifically linguistic semantic information that is difficult to access in olfaction. Regardless of whether accessing any type of semantic information, or only accessing linguistic semantic information is problematic, the difficulty in accessing information about recognized odors that cannot be named further illuminates the fragility of the connection between olfactory perception and language processes.

It can be speculated that we did not evolve a stronger connection between olfaction and language because language is not necessary for olfaction to perform its function. Olfactory information is not used for abstract problem solving. Instead, olfactory-guided behavior is mainly concerned with executing simple behaviors when an odor is encountered

(Herz 2001, 2005). In addition to this speculative evolutionary explanation, several neuroanatomic explanations for the poor connection between olfaction and language have been suggested. The lack of a thalamic relay in olfaction (Herz 2005), the fact that odor information is predominantly processed in the right hemisphere of the brain (for a review, see Royet and Plailly 2004) whereas language is predominantly expressed in the left hemisphere (Binder et al. 1997), and potential competition for computational resources (Lorig 1999) have all been suggested as contributors to our diminished capacity to name and talk about odors.

Whatever the reason for our inability to semantically process odor information is, it influences verbal reports about multimodal perceptions. Visual information always dominates when a verbal report is produced based on sensory information from different modalities. When visual and olfactory information are in conflict, the verbal report unfailingly reflects visual perception. This has been demonstrated in an experiment that set up a direct competition between conflicting visual and olfactory perceptions. Researchers asked students of the Faculty of Oenology of the University of Bordeaux to describe the taste of different wines. They tasted, in different sessions, a red wine (a cabernet-sauvignon/merlot) and a white wine (sémillon/sauvignon), as well as the same white wine, but with odorless red color added to it. The students described the taste of the white wine using words that are usually used to describe white wines. The red wine was described using words that are commonly found in descriptions of red wines. The interesting outcome of the experiments was the words that the students used to describe the taste of the wine that tasted like white wine but looked like red wine. The description of this wine was more similar to the description of the red wine than to the description of the white wine (Morrot et al. 2001). In other words, when visual information is available, the experts' description of wine taste is dominated by color rather than smell.

As part of the same study, the authors also analyzed the words used in thousands of wine tasting comments that they obtained from wine critics. They divided the tasting comments into those about white wines and those about red wines. What they found is that "the odors of a wine are, for the most part, represented by objects that have the color of the wine" (Morrot et al. 2001). Descriptors like "honey", "lemon", "grapefruit", "straw", and "banana" are often used to describe white wines, but never to describe red wines. On the other hand, the most common descriptors that are more frequently applied to red wines than to white wines are "cherry", "blackcurrant", "raspberry", "violet", and "redcurrant". Morrot and colleagues did not test the winemaking students whether they were capable of telling which of the three wines taste the same. It is likely that the students would have been able to distinguish between the red wine and the white wine with the red food color despite the color of the two wines being indiscriminable. Despite the inability to base verbal reports on olfactory perception, humans have an excellent sense of smell and perform very well in olfactory discrimination tasks (Bushdid et al. 2014). That experts can be tricked into verbally describing the taste of a white wine that is colored red as if they would describe a red wine is not a consequence of an underdeveloped sense of smell. It is a consequence of the dominance of vision over olfaction when it comes to producing a verbal report. Vision has a privileged connection to language processes and therefore has a stronger impact on verbal reports than conflicting information from other modalities such as olfaction.[2]

5.2 Olfaction and Evaluation

While olfaction has little impact on verbal reports about perception, it is often thought to play an important role in inducing and regulating certain emotions. Nabokov wrote, "Smells are surer than sights or sounds to make your heartstring crack." The same thought has been less poetically expressed by the psychologist Rachel Herz: "the sense of smell and emotional experience are fundamentally interconnected, bidirectionally communicative and functionally the same" (Herz 2007, p. 15). That smell and emotions are "functionally the same" means that there are striking similarities between how both odors and emotions motivate behaviors. It has been said that "More than any other sensory modality, olfaction is like emotion in attributing positive (appetitive) or negative (aversive) valence to the environment" (Soudry et al. 2011, p. 21). Humans use olfactory information mainly to evaluate food, locations, and other humans (Stevenson 2009). These evaluations result in changes in affective states and they are associated with highly adaptive behaviors.

Paradigmatic examples of the olfaction-emotion connection (which has been reviewed in detail before (Ehrlichman and Bastone 1992; Köster 2002; Herz 2007; Stevenson 2009)) are the influence of odors on emotions involved in romantic love and sexual arousal (Herz 2007; Stevenson 2009), and the close connection between olfaction and disgust (McBurney et al. 1977; Stevenson 2009; Stevenson et al. 2010). Disgust, love, fear, and sexual desire are examples of evaluative emotions. Only this type of simple evaluative emotion is closely connected to olfaction. Regulating more complex emotions, like jealousy or gratitude, requires an understanding of complex social relations and other people's intentions. Olfaction does not play a privileged role in the processing of this type of emotions.

The simple evaluative emotions that are closely connected to olfaction are often associated with physiological responses. Being disgusted increases the likelihood of shuddering, retching, and vomiting. Being sexually aroused increases heart rate and blood flow to the genitals. Emotions also are closely related to moods, which can be considered longer lasting states that increase the likelihood of specific emotions. Squeamish people are more easily and frequently disgusted and people with a high libido are more frequently and easily sexually aroused. It is an important and unresolved question what the relations between moods, emotions, and physiological responses are. The most notable dispute is whether, as proposed by William James, emotions are the perception of physiological responses. For the purpose at hand, it will not be necessary to answer these questions. Instead, I will limit myself to providing evidence for an exceptionally close connection between olfactory processing on the one side and evaluative emotions, moods, and physiological responses on the other side.

Olfaction as Inducer and Regulator of Evaluative Emotions

Odor perception is largely the perception of odor valence. Plato suggested that "pleasant" and "painful" are the only odor categories (Plato). More recently, multidimensional scaling techniques uncovered that valence is the most important perceptual dimension in olfaction (Haddad et al. 2008).

For colors and tones, valence is not an important perceptual dimension. When we are asked to arrange several odors in a one-dimensional space, we will likely order them at least in part according to their pleasantness. Colors, on the other hand, are more likely to be ordered from blue to red, and tones from low to high. This does not mean that all colors or all tones are equally pleasant. Very high tones are usually considered unpleasant and people tend not to like yellow-greenish colors. However, the difference in valence between the smell of rotting corpses and vanilla smell is larger than the difference in valence between yellow-green and your favorite color. Most people would rather live in an apartment in which the walls are painted in their least pleasant color than in an apartment that is filled with their least pleasant smell.

Further evidence for the close connection between olfaction and evaluative emotions is that emotional and physiological responses are more difficult to voluntarily modulate when they are odor-induced than when induced by other means. This shows that olfaction induces evaluative responses in an unmediated, direct, and automatic fashion. The smell of a preferred food is a potent inducer of salivation and subsequent consumption of the food. The smell of rotten corpses is a potent inducer of vomiting and subsequent behavioral odor avoidance. In comparison, pictures of food and pictures of rotten corpses are far less potent in inducing salivation or vomiting. Furthermore, the emotional and physiological responses induced by visual stimuli are easily modified by background information. The sight of a rotting corpse will not induce a strong affective response when the perceiver knows that it is an actor in make-up or a digital special effect in a movie. For smells, such background information is powerless to attenuate the affective response. The smell of decaying bodies can be recreated in the laboratory from synthetic molecules that have names like "putrescine" and "cadaverine". Exposing people to the synthetic corpse smell is likely to induce vomiting even when the subjects of the experiment have been told prior to the experiment that the smell they are about to perceive is a mixture of molecules that were synthesized in a factory, rather than the odor coming off rotten corpses. Overcoming visually induce physiological responses is much easier than overcoming odor-induced physiological responses. This difference shows that the connection between visual perception and emotional processes is much more flexible and fragile than the connection between emotion and olfaction.

All of the observations and experiments discussed above suggest that there is a privileged connection between olfaction and evaluative emotions. Skeptics will ask for an experiment in which the modalities are directly compared. However, comparisons between modalities are difficult because the results of the comparisons depend on the stimuli that were chosen for comparison (Ehrlichman and Bastone 1992). In one experiment, it was shown that odor stimuli elicited stronger affective responses than the corresponding visual stimuli. Subjects were asked to smell an odor, for example, the odor of freshly brewed coffee, or view a corresponding scene, for example, somebody pouring coffee from a pot into a cup. Then they were asked to write down "whatever immediately came to mind". Subjects wrote shorter reports in response to the olfactory stimulus than in response to the visual stimulus, indicating that verbal reports are dominated by visual input. However, the reports in response to the olfactory stimulus contained more affective words than the reports in response to the visual display (Hinton and Henley 1993).

Shared Neuroanatomy of Olfactory and Emotional Processes

Mechanistically, the close connection between olfaction and evaluative emotions can be explained in terms of neuroanatomy. There is large overlap between the brain regions that process emotions and smells (for a review, see Soudry et al. 2011). Much of the processing of emotions and olfactory information occurs in an evolutionary ancient brain structure called the limbic system.[3] Many of the brain structures in the limbic system play important roles both in the processing of olfactory information and in the processing of emotions. Consider, for example, the amygdala, an almond-shaped group of nuclei that is part of the limbic system. The amygdala is involved in the regulation of emotion (Aggleton et al. 2000; Salzman and Fusi 2010). Especially well studied is the role of the amygdala in regulating fear and aggression. In addition to this role, the amygdala also processes olfactory information. The amygdala receives strong direct input from the primary olfactory cortex, but very little direct input from the visual system (Zald and Pardo 1997; Gutiérrez-Castellanos et al. 2010; Pessoa and Adolphs 2010). In rats, around 40 % of the neurons in the amygdala are responsive to odors (Cain and Bindra 1972). Even more intriguingly, the connection between the primary olfactory cortex and the amygdala is bidirectional (Zald and Pardo 1997).

A second brain structure that is involved in both olfactory processing and the processing of emotions is the olfactory bulb. The olfactory bulb receives direct input from the olfactory sensory neurons. It is where the first steps of olfactory information processing happen. The olfactory bulb also plays a role in emotional regulation, which is surprising for a peripheral sensory structure that is only one synapse removed from sensory neurons. The olfactory bulb is so important for the processing of emotion that rodents in which the olfactory bulb has been removed surgically are an animal model for depression. The behavioral, endocrinological, and molecular changes seen in these animals are similar to those observed in patients with depression. Furthermore, these changes can be reversed by the same interventions that are used to treat patients suffering from depression, including antidepressants and electroconvulsive shock. The depression-like symptoms in mice without an olfactory bulb are not merely a response to the lack of olfactory input. Mice with an intact olfactory bulb in which olfactory input has been interrupted through other methods do not show depression-like symptoms. These results suggest that the olfactory bulb, which is the first and most important center of olfactory processing, also plays an important role in regulating emotions (for a review, see Song and Leonard 2005).

The part of the neocortex that processes olfactory information is the orbitofrontal cortex, which is located above the orbits in which the eyes are situated. The orbitofrontal cortex is only found in mammals (Gottfried 2007) and it is, unlike the visual cortex, not well connected to the frontal areas that are involved in semantic analysis (Price 2007). The role of the orbitofrontal cortex in olfactory processing is a matter of ongoing research. A lesion study of a single patient showed that brain injury that was largely limited to the right orbitofrontal cortex did completely abolish conscious processing of olfactory information. The patient's ability to modulate his sniffing behavior in response to olfactory stimuli was unaffected and he showed normal skin conductance responses to odors (Li et al. 2010). Based on this study, which is broadly consistent with previous studies of patients with orbitofrontal damage or lesions (see references in Li et al. 2010), it has been proposed that the orbitofrontal cortex is the neural correlate of olfactory consciousness. Others have suggested that the main role of the orbitofrontal cortex is to process the hedonic value of smells (Rolls et al. 2003). In addition to its role in olfactory perception, the orbitofrontal cortex also plays a key role in regulating affect, emotion, and motivation (Zald and Rauch 2008; Gottfried and Zelano 2011). The main role of the orbitofrontal cortex in this context seems to be to link reward to hedonic experience (Kringelbach 2005). Damage to the orbitofrontal cortex can lead to disinhibited behavior that can include gambling, swearing, drug addiction, and hypersexuality.

The amygdala, the olfactory bulb, and the orbitofrontal cortex are just three examples of brain structures that play important roles in olfactory processing as well as in the processing of emotions. Other structures within the limbic system show similar profiles. The large overlap of brain regions that process emotions and those that process olfactory information provide the mechanistic explanation for the privileged connection between olfaction and the processing of evaluative emotions.

5.3 Conclusion: Olfaction Is Well Connected to Emotional but Not to Linguistic Processing

The evidence presented in this chapter shows that olfaction has a strong impact on evaluative emotions, while our capacity to process olfactory information linguistically is very limited. This is not a new insight. Over 2000 years ago Plato wrote that odors "have no name and they have not many, or definite and simple kinds; but they are distinguished only as painful and pleasant" (Plato). Today we know that the reason for the privileged connection between olfaction and evaluative emotions is that the same neuronal networks in the brain that process olfactory information also process emotions. The connection between olfaction and emotions is presumably not the only privileged connection between a perceptual modality and a non-perceptual cognitive process. Vision seems to have a privileged connection to language processes. An analysis of proprioception, the sensing of the relative position and movement of body parts, would reveal a strong connection between proprioception and movement control. That one can find this type of modality-specific connections shows that sensory information from a given modality is made available only to those processes that can use the information for adaptive behaviors. The motor system needs to know the current angle between the forearm and upper arm, so that it can execute directed arm movements. The language system does not need to know the current elbow angle because being able to report the position of your forearm verbally does not convey strong adaptive advantages.

The sense of smell has evolved to be an evaluative rather than a descriptive sense. Olfactory information is used mainly to make decisions about rejecting or accepting food or mates (Stevenson 2009). Describing verbally the smell of spoiled meat is not crucial for survival; having a negative emotional response to spoiled meat that is stronger than hunger is crucial for an adaptive, odor-guided, behavioral response. The connection between olfaction and emotion is so close that Rachel Herz wondered "whether we would have emotions if we did not have a sense of smell; *I smell therefore I feel*?" (Herz 2007, p. 14). Herz's thoughts mirror those of Plato, who wondered whether we would have reason without vision and those of Michael Tomasello, who wondered whether we would have language without vision. Summarizing the different relations between perceptual modalities and cognitive processes, Trygg Engen wrote: "Functionally, smell may be to emotion what sight or hearing is to cognition" (Engen 1982, p. 3).

The philosophical impact of the heterogeneity in the connections between perceptual systems and non-perceptual systems is that epistemological accounts that are based on visual perception have to confront the fact that they cover only one, very specialized, form of perception. However, the more interesting point is metaphilosophical. The privileged connection between vision and language is the main reason why I felt that it was necessary to undertake the current research project to expose and correct the misguided ideas in the philosophy of perception that are based on the exclusive engagement with visual perception. The tool of philosophy is language and the connection between vision and language is stronger than the connection between other modalities and language, which gives vision privileged access to the minds of philosophers.

Notes

1. A similar proposal has been made by Michael Tomasello in his *Origins of Human Communication* Tomasello (2008). *Origins of Human Communication*. Cambridge, MIT Press. Tomasello argues that human communication evolved from joint attention and shared intentionality. Joint attention is the phenomenon of an individual attending to an object after observing that another individual attends to the object. When we come across a group of people looking out the window, we are likely to join them to find out what interesting thing is going on outside. Joining others' attention seems natural and does not require any conscious reasoning. However, being able to do that requires understanding what others perceive when their eyes are directed in a certain direction. This ability is sometimes referred to as "mindreading", because it requires inferring the content of another individual's mind in the absence of communication. This cognitively complex process is so sophisticated that it is rarely found in non-human animals. Joint attention in humans is only possible for visual attention. We can see what someone is looking at, but not hear what they are listening to, feel what they are touching, or smell what they are sniffing. Only through vision can one individual observe another individual in the process of perceiving. Tomasello et al. (2005). "Understanding and sharing intentions: The origins of cultural cognition." *Behavioral and Brain Sciences* **28**(5): 675–735.
2. Dominance of vision over information from other modalities during multimodal perception is often observed. A famous example is the ventriloquism effect. Although the voice attributed to the ventriloquist's dummy comes from the speaker's mouth, it is perceived as coming from the dummy's mouth because visually the dummy's mouth is perceived as moving whereas the speaker's mouth is not. What is special about the cases discussed here is that vision does not appear to change the olfactory perception as much as it specifically changes the verbal report.
3. This part of the brain is also known as "reptilian brain", because we share it with reptiles, or "rhinencephalon" (literally, "nose brain"), because it pro-

cesses smells. It is not a functionally unified system but rather a set of neighboring brain structures including the primary olfactory cortex, the limbic lobe, the hippocampus, and the amygdala.

References

Aggleton, J. P., Young, A. E., et al. (2000). The enigma of the amygdala: On its contribution to human emotion. In R. D. Lane & L. Nadel (Eds.), *Cognitie neuroscience of emotion* (pp. 106–128). Oxford: Oxford University Press.
Berlin, B., & Kay, P. (1969). *Basic color terms: Their universality and evolution.* Berkeley: University of California Press.
Binder, J. R., Frost, J. A., et al. (1997). Human brain language areas identified by functional magnetic resonance imaging. *The Journal of Neuroscience, 17*(1), 353–362.
Bushdid, C., Magnasco, M. O., et al. (2014). Humans can discriminate more than one trillion olfactory stimuli. *Science, 343*(6177), 1370–1372.
Cain, D. P., & Bindra, D. (1972). Responses of amygdala single units to odors in the rat. *Experimental Neurology, 35*(1), 98–100.
Chobor, K. L. (1992). A neurolinguistic perspective of the study of olfaction. In M. J. Serby & K. L. Chobor (Eds.), *Science of olfaction* (pp. 355–377). New York: Springer-Verlag.
Desor, J. A., & Beauchamp, G. K. (1974). The human capacity to transmit olfactory information. *Perception & Psychophysics, 16*(3), 551–556.
Ehrlichman, H., & Bastone, L. (1992). Olfaction and emotion. In M. J. Serby & K. L. Chobor (Eds.), *Science of olfaction* (pp. 410–438). New York: Springer-Verlag.
Engen, T. (1982). *The perception of odors.* New York: Academic.
Gilbert, A. N. (2008). *What the nose knows.* New York: Crown Publishers.
Gottfried, J. A. (2007). What can an orbitofrontal cortex-endowed animal do with smells? *Annals of the New York Academy of Sciences, 1121*(1), 102–120.
Gottfried, J. A., & Zelano, C. (2011). The value of identity: Olfactory notes on orbitofrontal cortex function. *Annals of the New York Academy of Sciences, 1239*(1), 138–148.
Gutiérrez-Castellanos, N., Martínez-Marcos, A., et al. (2010). Chemosensory function of the amygdala. *Vitamins and Hormones, 83*, 165–196.
Haddad, R., Khan, R., et al. (2008). A metric for odorant comparison. *Nature Methods, 5*(5), 425–429.

Herz, R. S. (2001). Ah, sweet skunk: Why we like or dislike what we smell. *Cerebrum, 3*(4), 31–47.

Herz, R. S. (2005). *The unique interaction between language and olfactory perception and cognition. Trends in experimental psychology research* (pp. 91–109). New York: Nova Science Publishers, Inc.

Herz, R. S. (2007). *The scent of desire*. New York: William Morrow.

Hinton, P. B., & Henley, T. B. (1993). Cognitive and affective components of stimuli produced in three modes. *Bulletin of the Psychonomic Society, 31*, 595–598.

Johnson, C. J., Paivio, A., et al. (1996). Cognitive components of picture naming. *Psychological Bulletin, 120*(1), 113–139.

Jönsson, F. U., Tchekhova, A., et al. (2005). A metamemory perspective on odor naming and identification. *Chemical Senses, 30*(4), 353–365.

Köster, E. P. (2002). The specific characteristics of the sense of smell. In C. Rouby, B. Schaal, D. Dubois, R. Gervais, & A. Holley (Eds.), *Olfaction, taste, and cognition* (pp. 27–44). Cambridge: Cambridge University Press.

Kringelbach, M. L. (2005). The orbitofrontal cortex: Linking reward to hedonic experience. *Nature Reviews Neuroscience, 6*, 691–702.

Lambon Ralph, M. A., Sage, K., et al. (2000). Classical anomia: A neuropsychological perspective on speech production. *Neuropsychologia, 38*(2), 186–202.

Levinson, S. C., & Majid, A. (2014). Differential ineffability and the senses. *Mind & Language, 29*(4), 407–427.

Li, W., Lopez, L., et al. (2010). Right orbitofrontal cortex mediates conscious olfactory perception. *Psychological Science, 21*(10), 1454–1463.

Lorig, T. S. (1999). On the similarity of odor and language perception. *Neuroscience & Biobehavioral Reviews, 23*(3), 391–398.

McBurney, D. H., Levine, J. M., et al. (1977). Psychophysical and social ratings of human body odor. *Personality and Social Psychology Bulletin, 3*, 135–138.

Morrot, G., Brochet, F., et al. (2001). The color of odors. *Brain and Language, 79*(2), 309–320.

Pessoa, L., & Adolphs, R. (2010). Emotion processing and the amygdala: From a 'low road' to 'many roads' of evaluating biological significance. *Nature Reviews Neuroscience, 11*(11), 773–783.

Price, J. L. (2007). Definition of the orbital cortex in relation to specific connections with limbic and visceral structures and other cortical regions. *Annals of the New York Academy of Sciences, 1121*(1), 54–71.

Rolls, E. T., Kringelbach, M. L., et al. (2003). Different representations of pleasant and unpleasant odors in the human brain. *European Journal of Neuroscience, 18*, 695–703.

Royet, J. P., & Plailly, J. (2004). Lateralization of olfactory processes. *Chemical Senses, 29*(8), 731–745.
Salzman, C. D., & Fusi, S. (2010). Emotion, cognition, and mental state representation in amygdala and prefrontal cortex. *Annual Review of Neuroscience, 33*(1), 173–202.
Schwartz, B. L., & Metcalfe, J. (2011). Tip-of-the-tongue (TOT) states: Retrieval, behavior, and experience. *Memory & Cognition, 39*(5), 737–749.
Song, C., & Leonard, B. E. (2005). The olfactory bulbectomised rat as a model of depression. *Neuroscience and Biobehavioral Reviews, 29*(4–5), 627–647.
Soudry, Y., Lemogne, C., et al. (2011). Olfactory system and emotion: Common substrates. *European Annals of Otorhinolaryngology, Head and Neck Diseases, 128*(1), 18–23.
Stevenson, R. J. (2009). Phenomenal and access consciousness in olfaction.
Stevenson, R. J., Oaten, M. J., et al. (2010). Children's response to adult disgust elicitors: Development and acquisition. *Developmental Psychology, 46*(1), 165–177.
Sulmont-Rosse, C. (2005). Odor naming methodology: Correct identification with multiple-choice versus repeatable identification in a free task. *Chemical Senses, 30*(1), 23–27.
Tomasello, M. (2008). *Origins of human communication.* Cambridge: MIT Press.
Tomasello, M., Carpenter, M., et al. (2005). Understanding and sharing intentions: The origins of cultural cognition. *Behavioral and Brain Sciences, 28*(5), 675.
Veramendi, M., Herencia, P., et al. (2013). Perfume odor categorization. *Journal of Sensory Studies, 28*, 76–89.
Zald, D. H., & Pardo, J. V. (1997). Emotion, olfaction, and the human amygdala: Amygdala activation during aversive olfactory stimulation. *Proceedings of the National Academy of Sciences, 94*(8), 4119–4124.
Zald, D. H., & Rauch, S. (2008). *The orbitofrontal cortex.* Oxford: Oxford University Press.

6

Modulation of Olfactory Perception

In the previous chapter, I have discussed the outputs of the olfactory system and how these outputs are used by cognitive processes. I argued that emotional processes' privileged access to olfactory information is a result of the interdependence and coevolution of perception and cognition. Emotional processes are one example of a non-perceptual system that evolved together with the sensory system from which it receives its input. Another example is the proprioceptor system, the sensory system that provides information about the relative position and movement of body parts. The proprioceptor system would not have evolved in the absence of a motor system that can use the information provided by it. In turn, the motor system would not have evolved the capacity to quickly adjust and correct ongoing motor patterns, if it had evolved in the absence of a sensory system that provides the information about the position of body parts that is necessary for guiding these corrective movements. A perceptual system only conveys an adaptive advantage in the presence of a non-perceptual system that can make use of the information that the perceptual system provides. Similarly, cognitive systems can only contribute to guiding adaptive behaviors when they receive information from sensory systems.

A. Keller, *Philosophy of Olfactory Perception*,
DOI 10.1007/978-3-319-33645-9_6

This chapter will complement the previous chapter. It is not about flow of information *from* olfactory perception to other processes, but about information flow *toward* the olfactory system. Recently, two questions about the influence of other parts of the mind on perceptual processes have received much attention. The first question is whether there is flow of information from cognitive processes to perceptual processes. This phenomenon has been termed “cognitive penetration of perception”. The second question is whether perceptual processes in different modalities can be part of the same network, a phenomenon known as “crossmodal perception”. The connections discussed in Chap. 5 are parts of networks in which information flows from the periphery to the center. In this chapter, networks in which information flows in the opposite direction or perpendicular to the flow from periphery to center will be discussed.

6.1 Cognitive Penetration of Perception

How the environment is perceived depends on the physical features of the environment and on the perceptual systems of the perceiver. In addition, it has been suggested that cognitive processes also affect perception. Examples of proposed effects of cognition on perception are that a steak smells different when one is hungry and when one has just eaten a steak (Gottfried 2007), or that wearing a heavy backpack makes hills look steeper (Bhalla and Proffitt 1999). Such effects of cognition on perception have been termed “cognitive penetration of perception” by Pylyshyn, who argued that early vision is cognitively impenetrable (Pylyshyn 1999).

There is a lively debate whether cognition has an effect on perception (Vetter and Newen 2014; Firestone and Scholl 2015). An often-cited argument for cognitive impenetrability of early visual perception is that several visual illusions resist cognitive influence. One example is the Müller-Lyer illusion: a straight line segment with arrowheads on both ends looks shorter than a line segment of the same length with two arrow tails on both ends. That this illusion persists after the subject understands that it is an illusion is shown by the fact that even after one measures the two line segments to confirm that they have the same length (Fig. 6.1a), the line segment with the arrowheads on its ends looks shorter than the

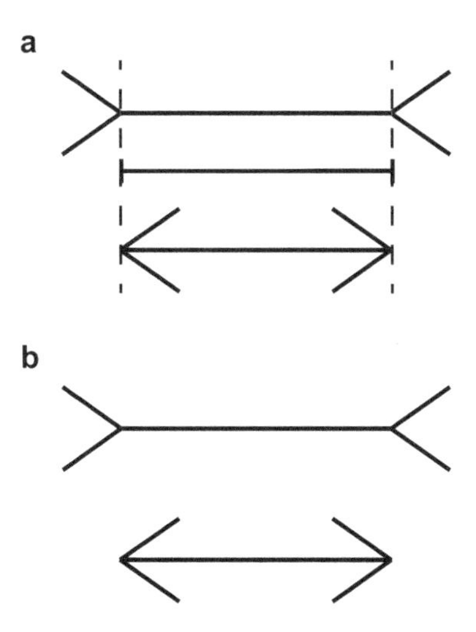

Fig. 6.1 The Müller-Lyer illusion, in which a line segment with two *arrowheads* on its ends looks shorter than a line segment of the same length with two *arrow tails* on its ends (**b**), persists even when the subject is made aware that the length difference is illusionary (**a**)

one with the arrow tails (Fig. 6.1b). Our knowledge about what we are seeing does not change how it looks to us.

Contrary to Pylyshyn, it has been argued that there are convincing examples of cognitive penetration in the visual system. According to Siegel (2006), for example, the perception of a pine tree changes when the perceiver learns to identify pine trees and to distinguish them from other types of trees. Similar examples can be given for other kinds of acquired expertise. To a trained pathologist, a tissue sample under the microscope may look different than it looked to her before her professional training. To a wine expert, wines may taste different than to somebody with no wine expertise. The problem with examples like this is that skeptics can simply deny that the perception changed. To address this concern, Macpherson discusses an interesting example of cognitive penetration of visual processing for which it is more difficult to deny the changes in perception because they are confirmed through performance-based psychophysics (Macpherson 2012). In an experimental setting, subjects were given shapes cut out of orange paper. Some of the shapes were the shapes of objects that are characteristically thought of as red (like a heart) whereas others were of the shapes of objects not characteristically red (like a mushroom). The task was to match each of the shapes to a colored background. The background color that the subjects matched to shapes of characteristically red objects was, on average, redder than the background color that was matched to shapes of objects that are not characteristically red. When the cutout was in the shape of a not characteristically red object, the subjects selected a more yellowish color (remember that the color of the cutouts was orange) (Delk and Fillenbaum 1965). This result provides strong evidence that the belief that a perceiver holds about the characteristic color of an object can influence the color it is perceived to have.

The discussion of cognitive penetration in the literature is mainly concerned with visual perception. As I have discussed in Chap. 5, the connections between perception and cognition evolved independently in each of the modalities and it is therefore not possible to generalize from visual perception to perception in general. Instead, each modality has to be investigated individually. In this section I will discuss cognitive penetration of olfactory perception.

Olfaction is a good modality for the study of cognitive penetration because introspection and behavioral experiments suggest that olfactory perception is strongly modulated by cognitive processes (for reviews, see Stevenson and Boakes 2003; Gottfried and Wu 2009; Yeshurun and Sobel 2010). An example of cognitive penetration in olfaction is the influence of verbal labels on perception. The pleasantness of the same stimulus is rated very differently when it is labeled “cheddar cheese” than when it is labeled “body odor”.[1] These differences in reported perception are also accompanied by differences in brain activity (de Araujo et al. 2005). Furthermore, anecdotal reports suggest that odor perception differs between different cultural groups. An impact of culture on perception indicates that beliefs and background information can penetrate perception. Unfortunately, the presumed cross-cultural differences in odor perception have not yet been studied systematically.

In the first part of this section I will argue that the question whether cognition penetrates perception presupposes that cognition and perception are two clearly separated processes and that, at least in olfaction, this is doubtful. In the second part, I will propose, following Lycan (2014), that instead of precisely delineating perceptual systems, cognitive penetration can be relativized to the processing stage of sensory information (Lycan

2014). I will then review the neural correlates for cognitive penetration at different stages of olfactory processing.

The Difficulty of Delineating Perception

Everybody, I presume, would agree that there is a difference between smelling and tasting food when hungry and smelling and tasting the same food after having eaten too much of it. The change in the experience is even more pronounced when there are strong positive or negative associations involved. Maybe a beer for breakfast is really the best cure for a hangover, but there is a pronounced difference between experiencing the flavor of beer pre- and post-alcohol poisoning. The disagreement over whether there is cognitive penetration of perception is a disagreement about the nature of the difference between these situations. Is it a difference in perception, or is it a difference in judgment about perception? Does alcohol smell and taste different post-alcohol poisoning, or does it taste the same but our judgment of that taste has changed? To answer this question, one has to decide first what counts as perception and what counts as judgment. Only if there is an independent reason to believe that perception and judgment are two separate processes does the question whether a given phenomenon is due to perception or due to judgment make sense.

How difficult it is to draw a line that separates perceptual processes from cognitive processes can be illustrated using Pylyshyn's article that introduced the term "cognitive penetration" (Pylyshyn 1999). The title promises to make a "case for cognitive impenetrability of visual perception". In the abstract, the part of the mind that is claimed to be impenetrable is then shrunk to "early vision". In the main text of the paper it is acknowledged that "what we see – the content of our phenomenological experience – is the world as we visually apprehend and know it: it is not the output of the [early] visual system itself". Many of the comments accompanying Pylyshyn's article discuss the difficulty of knowing the exact position of the border between "early vision", which according to Pylyshn is not influenced by cognition, and "later vision", which is.

The lack of agreement about the structure of the mind is, I think, the reason why there is disagreement about cognitive penetrability. Before one can decide whether a perceptual system receives cognitive input, one has to decide where perception stops and cognition starts. I am not aware of a principled way of doing that. Above, I briefly discussed the strategy of individuating the parts of the mind (that are then called "modules") depending on whether they satisfy certain sets of criteria. However, this strategy merely shifts the problem from deciding where the borders between the parts of the mind are to deciding what criteria should be used to draw the borders.

Admittedly, there are cases in which the distinction between perception and judgment is clear, especially in visual perception. A stick that is partially submerged in water is perceived as bent because of the different refraction of light in water and in air. Those familiar with this effect will judge the stick as being straight even when they perceive it as being bent. However, it is a mistake to conclude from the existence of seemingly unambiguous cases of perception and of judgment that there is a clear boundary between the two. Shaquille O'Neal (at 216 cm) is clearly tall whereas Danny DeVito (at 152 cm) is clearly short, but it does not follow from this that tall people and short people form two clearly separated groups.

An alternative strategy to define what counts as cognitive penetration of perception that does not require to first delineate perceptual processes is to define everything that alters the phenomenological experience associated with a stimulus as cognitive penetration. However, with such a notion of cognitive penetration, it is indisputable that it does exist. One uncontroversial example of such changes in perception is stress-induced analgesia, the phenomenon that pain is perceived to be less intense in stressful situations (Butler and Finn 2009).

Neural Correlates of Cognitive Penetration of Olfaction

Instead of delineating perceptual systems and then investigating whether the processing within them is influenced by cognitive processes, cognitive penetration can be relativized to processing stages (Lycan 2014). Using this approach, the most extreme version of cognitive penetration would be an influence of cognition on how sensory neurons translate physical stimuli into neuronal signals. If cognitive processes affect the function of sensory neurons, then there is no sensory information in our brains that has not been modified by cognition. Alternatively, cognitive information may only influence sensory processing at higher stages within the perceptual system. If this were the case, the information prior to the affected processing stage would be free of cognitive penetration. Later stages would be cognitively penetrated. This situation would be similar to Pylyshyn's proposal that visual perception can be divided into an early stage that is not penetrated by cognition and a later stage which is penetrated by cognition (Pylyshyn 1999). In this section, I will investigate the potential neural correlates for cognitive penetration at different levels of the olfactory system, starting with the olfactory sensory neurons.

Olfactory sensory neurons are the neurons in the nose that respond to odors. In humans, the responses of olfactory sensory neurons to the odors they encounter are not known to depend on cognitive processes. However, in other species strong influences of the perceiver's state on the activity of these sensory neurons have been described (see, e.g., the dependence of the sensitivity of olfactory sensory neurons on the perceivers' state in many insects discussed in Sect. 4.1. (Davis 1984; Root et al. 2011; Saveer et al. 2012)). The strategy behind such changes is to perceive only behaviorally relevant stimuli. If stimuli are relevant only in specific situations, then it is a good strategy to perceive them only during these situations. It does not appear that humans have developed such a system. We do not become smell blind to food odors after lunch.

The olfactory sensory neurons project to the olfactory bulb. In the olfactory bulb, information about the molecular structure of odorant molecules is processed (Shepherd et al. 2004). In addition to the input from olfactory sensory neurons, the olfactory bulb also receives information through massive and diverse centrifugal fibers. These fibers originate in brain areas including the olfactory cortex, the brain stem, and the basal forebrain (Matsutani and Yamamoto 2008). The information processing in the olfactory bulb is influenced by activity in all these diverse regions of the brain. Consistent with the anatomical findings, information processing in the olfactory bulb of rodents differs markedly depending on the behavior the animal is engaged in and its prior experiences (Kay and Laurent 1999). Processing in the olfactory bulb is, for example, modulated by prior experiences with the perceived odor (Kato et al. 2012). The massive feedback from higher brain areas to the olfactory bulb is the neural correlate of cognitive penetration in human olfaction. Through these connections, memories and mental states can influence how odors are perceived. It is here, one synapse away from the sensory neuron, that olfactory perception is modulated by cognitive input.

Overall, olfactory perception is perhaps more pliable than vision. The suggestion that a "grilled 40-oz. dry-aged porterhouse steak" smells different after one has just consumed such a steak (Gottfried 2007) is more plausible than the suggestion that the way the steak looks depends on whether the perceiver is hungry or satiated. This intuition can be denied. However, a striking fact about olfaction is that the olfactory bulb, the first processing station of olfactory information, receives massive input from central brain structures. Unless these structures are vestigial structures that have lost their function, olfaction is cognitively penetrated at a very early stage. To illustrate how early, the analogy to cognitive penetration one synapse away from the sensory neuron in the visual system would be cognitive penetration at the level of the horizontal cells in the retina.

6.2 Crossmodal Perception

In addition to cognitive processes, perception in a given modality can also be influenced by perception in other modalities. One of the most studied examples of crossmodal perception is the McGurk effect, which is the effect that watching the movements of a speaker's lips has on the perception of the sounds the speaker makes with their lips (McGurk and MacDonald 1976). Hearing the sound "ba" while watching lips that make the sound "ga" results in the perception of the sound "da". Demonstrations of the McGurk effect can be found online.

Many of the considerations for cognitive penetration also apply to crossmodal perception. Visual perception seems to be unusual among the modalities in that it is not commonly influenced by perception in other modalities. In the McGurk effect, visual input determines what we hear, not the other way around. Intuitively, it does not feel that what we see is modulated by what we smell or hear. Instead, what we see influences what we perceive in the other modalities and the non-visual modalities influence each other. As with cognitive penetration of perception, whether something counts as intermodal penetration of perception depends on the delineation of perception. One could argue that in the

McGurk effect, what changes is not the auditory perception itself, but the interpretation of what is perceived. Like with cognitive penetration of perception, it is therefore useful to relativize the crosstalk between perceptual modalities to processing stages.

Here, I will first discuss flavor perception as a paradigm case of crossmodal perception and then investigate the neural correlates for crossmodal perception.

Flavor Perception

One of the most complex cases of crossmodal perception is the perception of flavor. "Flavor" is the scientific term for what is commonly called "taste". Scientists use the word "taste" for the perception that is mediated by the taste buds on the tongue whereas they use the word "flavor" for the multisensory percepts elicited by food in the oral cavity. The chemical senses like smell and taste together with the sensing of textures and temperatures produce the perception of flavor, which has been called a "multisensory modality" (Taylor and Roberts 2004; Shepherd 2011; Small and Green 2012).

Taste is the main contributor to flavor. Humans perceive five basic taste qualities: sweetness, sourness, saltiness, bitterness, and umami (a savory taste). Recent research suggests that we also have dedicated sensory neurons for other tastants, like fat (Laugerette et al. 2007) and calcium (Tordoff et al. 2012). Flavor nuances are added to these basic tastes through the contribution of olfaction. When perceived through taste alone, all Jelly Beans have the same flavor: sweet. However, when olfaction is added, some Jelly Beans have coconut flavor whereas others have popcorn flavor. One can experience this difference by pinching the nose shut with two fingers, then starting to chew a Jelly Bean and then opening the nose. The shift in flavor from sweet to sweet pineapple is due to the odorous molecules released by the Jelly Beans that can reach the olfactory epithelium only when air flows from the mouth to the nose through the pharynx (a process known as retronasal olfaction (Bojanowski and Hummel 2012)). In addition to taste and olfaction, chemesthesis also contributes to perceived flavors. Chemesthesis is the sensitivity to chemicals of areas of the skin or mucous membranes. The mucous membrane of the oral cavity contains many chemesthesis-mediating neurons (Green 2004). The "hotness" of chili peppers and the "coolness" of menthol do not involve temperature perception, but chemesthetic perception of molecules that are sensed by temperature sensors. Similarly, the "tingling" induced by carbonated drinks is not touch but chemesthesis (Green 2004). In addition to the chemical senses, touch, temperature, and other senses also contribute to flavor perception.

"Flavor" is not just a term for the simultaneous occurrence of perceptions in a variety of modalities. Instead, the different modalities involved in flavor perception influence each other. For reviews of the crossmodal interactions in flavor perception, see Taylor and Roberts (2004) and Shepherd (2011). An example of intermodal modulation of olfactory perception during flavor perception is that combining a tastant with an odorant can enhance the perceived intensity of the odorant (Green et al. 2011) as well as its detectability (Dalton et al. 2000).

Neural Correlate of Crossmodal Perception

The principal evidence that perception in one modality influences simultaneous perception in another modality comes from introspection and behavioral studies. Such evidence is easily disputed. In fact, some of the top comments on the websites hosting the demonstration video of the McGurk effect are from people reporting to be immune to the influence of seeing lip movements on their auditory perception. That there is indeed large variability in the susceptibility to the effect has been confirmed in laboratory studies (Nath and Beauchamp 2012). Because there is no easy way of adjudicating between different introspections, it is important to investigate whether there are neuronal structures that could mediate crossmodal perception.

Neuroanatomical investigation can reveal whether the required wiring for crossmodal penetration of perception is present. There are two complications to evaluating whether a neuronal structure should count as a potential neuronal correlate of crossmodal perception. The first complication is that interactions between two modalities may be direct or indirect. In the discussion of cognitive penetration of perception, I have already mentioned the massive feedback from central brain structures to the olfactory system. This feedback could indirectly carry information about perception in other modalities, thereby providing the basis for crossmodal perception. The second complication is that, as with cognitive penetration, the interaction can be at different levels of processing. A neuronal connection between photoreceptors and olfactory sensory neurons would be a very clear indication of information exchange between vision and olfaction. But what about a connection between two modalities at higher levels of sensory processing?

A review of the neuroanatomy of the neocortex has concluded that there are many connections between neurons processing information from different modalities. In fact, the interactions are so ubiquitous that it is difficult to identify individual modalities (Shimojo and Shams 2001). However, anatomical connections between the olfactory system and the other sensory systems at lower levels of processing than the neocortex are sparse. The lowest levels of processing at which visual, gustatory, and olfactory information are thought to converge are the orbitofrontal cortex and the amygdala (Rolls and Baylis 1994; Rolls et al. 2009). Despite the lack of direct anatomical connections between olfaction and other sensory systems in the periphery, it has been shown that in rodents piriform cortex neurons are influenced by gustatory stimuli (Maier et al. 2012). If these findings generalize to the human brain, the feedback neurons that are the substrate of cognitive penetration are also the best candidate for the neural correlate of the strong crossmodal influences on olfactory perception.

6.3 Conclusion: Olfaction Is Bidirectionally Connected to Other Modalities and Cognitive Processes

In this chapter, I have described the inputs that olfaction receives from cognitive processes and from other perceptual modalities. Olfaction, probably more so than other perceptual modalities, is modulated by both types of input. Comparing the facts presented here about olfaction with the corresponding facts about vision shows that questions about cognitive penetration and crossmodal perception cannot be answered generally.

Instead, these questions have to be addressed separately for each modality and for each stage of processing. The reason for this lack of shared principles in how perceptual processes are connected to other processes is that each modality has been shaped by evolution through natural selection according to its specific functions and constraints. Olfaction followed its own evolutionary trajectory and found its own unique, idiosyncratic solutions to the problem it evolved in response to.

Considering the topics of cognitive penetration and crossmodal perception in olfaction leads to very different results than considering these topics in vision. The differences between modalities is a reminder that our mind has a complex structure and that there is no overarching design principle. The lack of order and homogeneity in our mind prompted Gary Marcus in his book *Kluge: The Haphazard Construction of the Mind* to call the mind a "kluge" (Marcus 2008). A "kluge" is a solution to a problem that is effective yet inelegant and clumsy. In the absence of generalizable rules about the structure of the mind, the mind needs to be described to be understood.

Note

1. Part of this effect is very likely due to the subjects' desire to conform to social norms. Subjects may rate anything labeled "body odor" as unpleasant to not appear unhygienic and anything labeled "cheese" as pleasant to appear sophisticated. One could also argue that pleasantness is not perceived, but judged. Firestone and Scholl (2015). "Cognition does not affect perception: Evaluating the evidence for 'top-down' effects." *Behavioral and Brain Sciences*. However, as I have discussed above, valence is a perceptual dimension in olfaction.

References

Bhalla, M., & Proffitt, D. R. (1999). Visual-motor recalibration in geographical slant perception. *Journal of Experimental Psychology. Human Perception and Performance, 25*, 1076–1096.

Bojanowski, V., & Hummel, T. (2012). Retronasal perception of odors. *Physiology & Behavior, 107*(4), 484–487.

Butler, R. K., & Finn, D. P. (2009). Stress-induced analgesia. *Progress in Neurobiology, 88*(3), 184–202.
Dalton, P., Doolittle, N., et al. (2000). The merging of the senses: Integration of subthreshold taste and smell. *Nature Neuroscience, 3*(5), 431–432.
Davis, E. E. (1984). Regulation of sensitivity in the peripheral chemoreceptor systems for host-seeking behavior by a hemolymph-borne factor in Aedes aegypti. *Journal of Insect Physiology, 30*, 179–183.
de Araujo, I. E., Rolls, E. T., et al. (2005). Cognitive modulation of olfactory processing. *Neuron, 46*(4), 671–679.
Delk, J. L., & Fillenbaum, S. (1965). Differences in perceived colour as a function of characteristic color. *The American Journal of Psychology, 78*(2), 290–293.
Firestone, C., & Scholl, B. J. (2015). Cognition does not affect perception: Evaluating the evidence for 'top-down' effects. *Behavioral and Brain Sciences, 20, 1–77.*
Gottfried, J. A. (2007). What can an orbitofrontal cortex-endowed animal do with smells? *Annals of the New York Academy of Sciences, 1121*(1), 102–120.
Gottfried, J. A., & Wu, K. N. (2009). Perceptual and neural pliability of odor objects. *Annals of the New York Academy of Sciences, 1170*(1), 324–332.
Green, B. G. (2004). Oral chemesthesis: An integral component of flavour. In A. J. Taylor & D. D. Roberts (Eds.), *Flavor perception* (pp. 151–171). Oxford: Blackwell Publishing.
Green, B. G., Nachtigal, D., et al. (2011). Enhancement of retronasal odors by taste. *Chemical Senses, 37*(1), 77–86.
Kay, L. M. & Laurent, G. (1999). Odor- and context-dependent modulation of mitral cell activity in behaving rats. *Nature Neuroscience, 2*(11), 1003–1009.
Kato, Hiroyuki K., Monica W. Chu, et al. (2012). Dynamic sensory representations in the olfactory bulb: modulation by wakefulness and experience. *Neuron, 76*(5), 962–975.
Laugerette, F., Gaillard, D., et al. (2007). Do we taste fat? *Biochimie, 89*(2), 265–269.
Lycan, W. G. (2014). What does vision represent? Does Perception have Content? B. Brogaard. Oxford: Oxford University Press, 311–328.
Macpherson, F. (2012). Cognitive penetration of colour experience: Rethinking the issue in light of an indirect mechanism. *Philosophy and Phenomenological Research, 84*(1), 24–62.
Maier, J. X., Wachowiak, M., et al. (2012). Chemosensory convergence on primary olfactory cortex. *Journal of Neuroscience, 32*(48), 17037–17047.
Marcus, G. (2008). *Kluge: The haphazard construction of the mind.* Boston: Houghton Mifflin Company.
Matsutani, S. and Yamamoto, N. (2008). Centrifugal innervation of the mammalian olfactory bulb. *Anatomical Science International 83*(4), 218–227.

McGurk, H., & MacDonald, J. (1976). Hearing lips and seeing voices. *Nature, 264*(5588), 746–748.
Nath, A. R., & Beauchamp, M. S. (2012). A neural basis for interindividual differences in the McGurk effect, a multisensory speech illusion. *NeuroImage, 59*(1), 781–787.
Pylyshyn, Z. (1999). Is vision continuous with cognition? The case for cognitive impenetrability of visual perception. *Behavioral and Brain Sciences, 22*, 341–423.
Rolls, E. T., & Baylis, L. L. (1994). Gustatory, olfactory, and visual convergence within the primate orbitofrontal cortex. *Journal of Neuroscience, 14*(9), 5437–5452.
Rolls, E. T., Critchley, H. D., et al. (2009). The representation of information about taste and odor in the orbitofrontal cortex. *Chemosensory Perception, 3*(1), 16–33.
Root, C. M., Ko, K. I., et al. (2011). Presynaptic facilitation by neuropeptide signaling mediates odor-driven food search. *Cell, 145*(1), 133–144.
Saveer, A. M., Kromann, S. H., et al. (2012). Floral to green: Mating switches moth olfactory coding and preference. *Proceedings of the Royal Society B: Biological Sciences, 279*(1737), 2314–2322.
Shepherd, G. M. (2011). *Neurogastronomy: How the brain creates flavor and why it matters*. New York: Columbia University Press.
Shepherd, G. M., & Chen, W. R. et al. (2004). Olfactory Bulb. The Synaptic Organization of the Brain. G. M. Shepherd. New York, Oxford University Press: 165–216.
Shimojo, S., & Shams, L. (2001). Sensory modalities are not separate modalities: Plasticity and interactions. *Current Opinion in Neurobiology, 11*(4), 505–509.
Siegel, S. (2006). Which properties are represented in perception? In T. Szabó Gendler & J. Hawthorne (Eds.), *Perceptual experience* (pp. 481–503). Oxford: Oxford University Press.
Small, D. M., & Green, B. G. (2012). A proposed model of a flavor modality. In M. M. Murray & M. T. Wallace (Eds.), *The neural bases of multisensory processes*. Boca Raton: CRC Press.
Stevenson, R. J., & Boakes, R. A. (2003). A mnemonic theory of odor perception. *Psychological Review, 110*(2), 340–364.
Taylor, A. J., & Roberts, D. D. (Eds.). (2004). *Flavor perception*. Oxford: Blackwell Publishing.
Tordoff, M. G., Alarcón, L. K., et al. (2012). T1R3: A human calcium taste receptor. *Scientific Reports, 2*, 496.
Vetter, P., & Newen, A. (2014). Varieties of cognitive penetration in visual perception. *Consciousness and Cognition, 27*, 62–75.
Yeshurun, Y., & Sobel, N. (2010). An odor is not worth a thousand words: From multidimensional odors to unidimensional odor objects. *Annual Review of Psychology, 61*(1), 219–241.

Part 4

Consciousness

So far, I have discussed perception largely without reference to consciousness.[1] However, a complete account of perception has to acknowledge that there is a fundamental difference between perception without consciousness and perception that is accompanied by the conscious experience of what is perceived. The main result of consciousness research so far has been that consciousness is a property of brain activity. There is no consciousness in the absence of brain activity and many manipulations of the activity of neuronal networks in the brain, like psychoactive drugs, brain damage, or electrical activation of neurons (e.g., stimulation of the auditory nerve through microelectrodes in cochlear implants), are known to change conscious perception. However, not all brain activity is conscious. The current debates in consciousness research are mainly about the difference between conscious and non-conscious brain activity in terms of underlying mechanisms and function.

Before conscious perception is discussed, it has to be clarified what the difference between conscious and non-conscious perception is. This is especially important because the term "consciousness" is ambiguous and several theorists have proposed subdivisions to differentiate between different notions of consciousness. Ned Block, for example, divides consciousness into phenomenal and access consciousness (Block 1997). Phenomenal consciousness is the qualitative nature of an experience ("What it is like to smell coffee."), whereas access consciousness is the ability to report (not necessarily verbally) an experience. Gerald Edelman divides consciousness into primary consciousness and higher-order consciousness (Edelman 2003). Primary consciousness is what William James called "specious present" and Edelman calls "the remembered present". Higher-order consciousness is more complex and restricted to animals with semantic abilities. Thomas Metzinger suggests a constraint satisfaction approach according to which the degree of phenomenality depends on how many of 11 constraints are satisfied by neural representations. Based on this approach he proposes four different notions of consciousness: minimal consciousness, differentiated consciousness, subjective consciousness, and cognitive, subjective consciousness. According to Metzinger, a system that possesses only minimal consciousness "would be frozen in an eternal Now, and the world appearing to this organism would be devoid of all internal structure" (Metzinger 2003, p. 204). Finally, Antonio Damasio distinguishes between core consciousness and the more sophisticated extended consciousness which is usually found in conjunction with complex language skills (Damasio 1999). I will limit the discussion here to the least complex form of consciousness. Depending on whose taxonomy one follows, this would be phenomenal, primary, minimal, core consciousness. The reason for focusing on these least complex forms is the hope that they are easier to understand than more complex forms and that it is therefore a good research strategy to start with phenomena of lower complexity.

Most discussions of consciousness employ a much broader notion of consciousness. One prominent example is the notion of consciousness that is used by proponents of the global workspace theory of consciousness. Within this framework, the paradigm example for non-conscious olfactory perception is the tip-of-the-nose-phenomenon (Baars 2013). The tip-of-the-nose phenomenon is the familiar situation of smelling a common odor, but being unable to name it. According to the notion of consciousness employed in the global workspace framework, somebody who reports smelling an odor she is unable to name is considered to perceive the odor non-consciously. In contrast, according to the notion of consciousness that I use here, she would be considered to perceive the odor consciously, even though she is unable to name it. In other words, the notion of consciousness employed here is extremely inclusive. Conscious processing is considered to take place whenever the subject has brain activities for which there is *something it is like for the organism to execute them*, to adapt an expression from Thomas Nagel (1974). The contrast group of non-conscious brain activities are those brain activities for which there is nothing it is like for the organism to execute them, just like there is nothing it is like for the organism to filter blood in the kidney.

A consequence of my focus on the least complex forms of consciousness is that it is difficult to relate the topics discussed here to much of the literature on consciousness which concerns more sophisticated forms of consciousness. Another difference between the discussion here and most other works on consciousness is that, in keeping with the topic of this book, I will focus on olfaction instead of vision. The olfactory system is an ideal model for investigating the mechanisms and functions of consciousness. The advantage of olfaction for consciousness studies is that in olfaction information processing of the same stimulus often switches between conscious and non-conscious. The air in the nasal cavity almost always contains enough odorous molecules to activate olfactory sensory neurons. Activation of olfactory sensory neurons will result in neuronal activity in the olfactory bulb and the olfactory cortex. Most of the time, this odor-induced brain activity is not conscious. We are usually not aware of the presence of odorants in the air surrounding us. Ezequiel Morsella and colleagues call the common lack of olfactory conscious experience "experiential nothingness". They argue convincingly that the absence of olfactory experience is more similar to the phenomenology of the blind spot than to the visual experience of darkness (Morsella et al. 2010). However, occasionally the same odor stimuli that are usually processed non-consciously do induce conscious brain activities. These switches from processing the same information non-consciously to processing it consciously can be studied to elucidate the underlying mechanisms and the function of conscious processing.

How conscious and non-conscious brain activities differ in terms of their underlying mechanisms and functions are the main questions of consciousness research. I will discuss mechanisms with a special focus on attention in Chap. 7. In Chap. 8, I will argue that it is the function of conscious perception to facilitate decision making in situations with many behavioral options.

Note

1. With some exceptions like the discussion of phenomenological presence as a criterion for objecthood in Sect. 3.3.

References

Baars, B. J. (2013). Multiple sources of conscious odor integration and propagation in olfactory cortex. *Frontiers in Psychology*, 4, 930.

Block, N. (1997). On a confusion about a function of consciousness. In *The nature of consciousness* (pp. 375–416). Cambridge: MIT Press.

Damasio, A. R. (1999). *The feeling of what happens: Body, emotion and the making of consciousness*. London: Heinemann.

Edelman, G. M. (2003). Naturalizing consciousness: A theoretical framework. *Proceedings of the National Academy of Sciences, 100*(9), 5520–5524.

Metzinger, T. (2003). *Being no one: The self-model theory of subjectivity*. Cambridge: MIT Press.

Morsella, E., Krieger, S. C., et al. (2010). Minimal neuroanatomy for a conscious brain: Homing in on the networks constituting consciousness. *Neural Networks, 23*(1), 14–15.

Nagel, T. (1974). What is it like to be a bat? *The Philosophical Review, 83*(4), 435–450.

7

Mechanisms of Consciousness

A subset of neuronal processes is conscious while most neuronal processes are not conscious. Understanding the differences between these types of processes is an important first step toward elucidating the mechanisms underlying consciousness. Understanding the mechanisms of conscious information processing is a very ambitious goal, yet some theorists argue that reaching this goal would not be satisfying. They suggest that even a complete understanding of the necessary and sufficient conditions for conscious brain activities would leave an explanatory gap (Levine 1983). Closing this explanatory gap has been called the "hard problem" (Chalmers 1995). This "hard problem" is defined by its impenetrability by science; it "persists even when the performance of all the relevant functions is explained" (Chalmers 1995). Unlike many other intuitions that concern hypothetical situations, the intuition that the explanatory gap will remain after a full understanding of conscious brain processes has been achieved has the advantage that we can work toward turning the hypothetical situation into an actual situation. Once we have a theory that allows accurate predictions about whether a process is conscious or not, we will see whether an explanatory gap remains. The explanatory gap is philosophically important because it has been suggested that an explanatory gap

A. Keller, *Philosophy of Olfactory Perception*,
DOI 10.1007/978-3-319-33645-9_7

indicates the existence of an ontological gap: if consciousness cannot be satisfyingly explained in terms of physical processes, then consciousness is not a physical process (see Chalmers 2006). The presumption of my discussion of consciousness is that consciousness is a physical process.

Despite the increasing research activity in the field, there are no convincing theories about the mechanisms of conscious neuronal processes. The discussions remain highly speculative. In this chapter, I will first discuss the methodology of identifying the mechanisms of conscious neuronal processing. My suggestion is that it is an important first step to identify phenomena and processes that correlate well with conscious processing in the brain. Correlation does not imply causation but it is unlikely that a process that does not correlate well with conscious perception is mechanistically involved in it. In the second part of the chapter, I will then discuss a cognitive process, attention, that correlates very closely with conscious processes.

7.1 Identifying the Mechanisms of Consciousness

There is no shortness of proposals about what the mechanisms of consciousness are. However, the proposed theories are rarely evaluated. Consciousness research is at a stage at which theory development is valued more than theory testing. An objective way of evaluating mechanistic theories of consciousness is to test how well the proposed mechanism correlates with the occurrence of conscious processes.

Evaluating Mechanistic Theories of Consciousness

A problem with evaluating predictive mechanistic theories of consciousness is that the presence or absence of consciousness is difficult to establish. Let us assume that a theory predicts that a process that happens inside of computers, like the integration of information, is conscious. Since there is no agreed-upon way to verify or falsify whether processes that occur inside a computer are conscious, this prediction cannot be used to evaluate the theory. The same problem of evaluating predictions is encountered when the predictions concern processes in animals. Whatever a theory predicts about the conscious states of octopuses can neither be used in support of the theory nor as an argument against it. Even in humans, there can be disagreement about whether a certain process is conscious or not. A further complication is that consciousness is not an all-or-none phenomenon. Several experiments suggest that consciousness is gradual (Kouider et al. 2010). Furthermore, when subjects are directly asked about their visual experiences, they tend to report them as being graded rather than as completely conscious or completely non-conscious (Overgaard et al. 2008). Despite these complications, in some cases there is wide agreement about what a mechanistic theory of consciousness should predict. There is overwhelming evidence that processes in the human liver or retina are much less likely to be conscious than processes in the human visual cortex. A simple test whether a process is involved in conscious processing is therefore to test whether it correlates stronger with activities in the cortex than in the liver and retina. Surprisingly, many mechanistic theories fail this test.

There are many differences between the liver, retina, and cortex. However, these differences are mainly differences in high-level organization at the cellular and molecular level. At the atomic and subatomic level, there are no known differences between different organs, which is why mechanistic theories of consciousness that evoke quantum mechanics (Koch and Hepp 2006; Atmanspacher 2011) fail to predict the absence of conscious liver processes. Other theories evoke general features of brain processes like synchronous firing of neurons (O'Brien and Opie 1999), the oscillation of such synchronous oscillations (Uhlhaas et al. 2009; Singer 2011), or the integration of information (Tononi 2008). These features of brain activity do not correlate with the unconscious liver processes but they generally fail to distinguish between the unconscious processes in the retina and the conscious processes in the cortex.

Necessary Conditions for Conscious Processing

The pitfall of much writing on the mechanisms of consciousness is that much more effort is put into showing that a certain phenomenon is present when conscious processing occurs than into showing that the phenomenon is absent when no conscious processing occurs. Consequently, many proposed mechanisms of conscious processing are merely necessary conditions. Theories like the synchronous oscillation theory and the information integration theory propose that basic features of simple neuronal networks play a role in conscious information processing. Even trivially simple networks, when they are large enough, integrate enough information that the information integration theory would judge them to be conscious (Seth et al. 2006). Similarly, many small, randomly connected networks show oscillatory activity (Pham et al. 1998). Even groups of neurons that grow in a petri dish will spontaneously form networks that sustain oscillatory activity (Muramoto et al. 1993; Idelson et al. 2010). In brains, oscillations are found in a wide variety of animals while they are awake as well as during slow-wave sleep and anesthesia (Steriade et al. 1996a, b; Vanderwolf 2000). In human brains many regions in which information is not processed consciously, like the retina (Neuenschwander and Singer 1996) and the hippocampus (Bragin et al. 1995; Colgin and Moser 2010), also show pronounced oscillations. Finally, the synchronous firing of neurons in the visual system persists in the resting state, in the absence of a stimulus (Brunet et al. 2014).[1]

Because of the ubiquity of synchronous oscillations and information integration in the brain, there is no close correlation between these phenomena and conscious processes. The corresponding theories therefore fail to make correct predictions about which brain processes are conscious and which are not. However, this does not mean that synchronous oscillations are not involved in conscious processes. It only means that they are not sufficient for consciousness. Oscillations may be necessary for conscious brain activities, in the same way in which living cells and neuronal activity are necessary for conscious information processing. Identifying necessary conditions is progress toward understanding a mechanism. However, Information integration and synchronous oscillations do not correlate very well with conscious processes, which makes it unlikely that they play a central role in what differentiates conscious from non-conscious brain activities. Unfortunately, very little is known about processing in the human brain at the level of neurons and neuronal networks. There are therefore, to my knowledge, no phenomena known at this level of description that correlate strongly with conscious processing. However, at a higher level of description, the level of cognitive processes, there are phenomena that correlate very closely with conscious processing. One of the most interesting examples is attention.

7.2 Attention

Attention is strongly correlated with conscious processes. The ticking of a clock or the touch of your tongue against the roof of your mouth are only processed consciously when attended to. The type of attentional shift that is involved in these examples is the allocation of attention toward a modality. Attending to audition results in conscious processing of background noises whereas shifting attention toward (passive) touch leads to conscious processing of all the contact points our skin has with the environment. The attentional shift between modalities is the most basic form of attention allocation. A simple model to study the effects of attention is the attentional shift toward the olfactory modality. Attention can be shifted to olfaction in the same way it can be shifted between other modalities. The attentional shift between audition and vision has been studied in detail (Spence and Driver 1997). Because of the potential applications in the management of chronic pain, the shifts of attention to and from nociceptive stimuli are also well understood (Eccleston 1995). Attentional shifts to olfaction are less well-studied, but several psychophysical studies have shown that attention can be shifted toward olfaction. Evidence for such a shift is that attending to olfaction decreases the response time to odors (Spence et al. 2001). In addition to behavioral effects of attending to smells, physiological effects have been described (for example (Krauel et al. 1998)) and more recently attention to odors has been shown to alter both behavioral responses to odors and odor-induced patterns of brain activity (e.g., Zelano et al. 2005). Together these data show that attending to olfaction is possible in much the same way in which we may attend to vision or audition.

The relation between attention and consciousness has often been discussed in terms of necessity and sufficiency. It has, for example, been suggested that attention is both necessary and sufficient for consciousness (Prinz 2012). Others have cited counterexamples both against the necessity of attention (van Boxtel et al. 2010) and against the sufficiency (Norman et al. 2013). It has also been suggested that the causal relationship between attention and consciousness is reversed and that consciousness is necessary for attention (Mole 2008). Alternatively, the relation between attention and consciousness could be more complicated. According to Michael Graziano's attention schema theory, awareness is a description of attention (Graziano 2013). I will limit myself here to arguing that there is a close correlation between attention and conscious processing in olfaction. Sufficiency and necessity are complicated concepts that apply only in situations that are defined in great detail. Is a functioning car with a full tank of gas sufficient to drive from Boston to Philadelphia, or is it not sufficient because you also need roads, bridges, arms, eyes, a driver's license, oxygen to breath, gravity, and so on?

Correlation Between Attention to Olfaction and Olfactory Consciousness

Olfaction researchers seem to agree that there is a very close connection between attention and olfactory consciousness. Sela and Sobel for example write that "olfactory stimuli are less prone to attract attention, and therefore humans have poor awareness to the olfactory environment" (Sela and Sobel 2010). Similarly, Köster and colleagues write that "conscious odor perception normally only occurs in situations where attention is demanded" (Köster et al. 2014, p. 1). The widespread view that olfactory consciousness is normally dependent on attention is supported by both observational and experimental evidence.

Everyday olfactory experiences show the close connection. We inhale air that contains odors with almost every breath; yet olfactory experiences are very rare (compared, for example, with visual experiences). This shows that an additional cognitive factor is necessary to switch from non-conscious processing of olfactory information to conscious processing of olfactory information. Attending to the olfactory modality usually results in a conscious olfactory experience (try it now!). The importance of the role of attention for olfactory experiences is further illustrated by the fact that people are very often wrong in their judgments about changes in their own olfactory abilities or in the odorous environment. The natural assumption, when a person's conscious olfactory experiences change systematically is that either the stimuli or the sensory apparatus has changed. However, in numerous well-studied situations, this is not the case. Instead, the change in conscious olfactory experience is entirely caused by a change in attention to the olfactory modality. Increased consciousness of smells due to increased attention to smells is seen during pregnancy, but also in people in which the cause for the change is not known. The vast majority of pregnant women report that their olfactory sensitivity is enhanced during pregnancy. However, studies have shown that the ability to detect odors at low concentrations does not change during pregnancy (Cameron 2007; Doty and Cameron 2009). Instead, the systematic differences in conscious olfactory experiences are caused by attentional factors. The involuntary increase in attention to odors during pregnancy is probably an adaptive response to the vulnerability of the fetus to environmental poisons and spoiled food. These attentional changes result in a radically altered olfactory conscious experience.

Observations like this are also supported by the results of experiments that have revealed a stunning failure of subjects to become conscious of unattended odor stimuli (Degel and Koester 1999). In one study (Lorig 1992), in which the influence of odors on the appeal of pictures was studied, only three out of 93 subjects became aware of the odor manipulation whereas several other subjects reported a perceived (although nonexistent) change in luminance.

Often, we do not consciously smell anything despite the presence of an olfactory stimulus until we allocate attention to olfaction, triggered, for example, by somebody exclaiming, "What's that smell?". However, there are also numerous examples of smells that are perceived consciously although we have not shifted attention to the olfactory modality. Ethyl mercaptan, which is added to natural gas to facilitate the locating of gas leaks, for example, is often perceived by people who do not attend to smells at all. Ethyl mercaptan is added to the odorless natural gas at 57,000 times the concentration at which it can be detected when attended to (Sela and Sobel 2010). It is an extremely strong stimulus that draws attention to olfaction. This is another case of conscious perception of smells correlating with attending to olfaction.

My proposal about attention is that it is best thought of as a resource that can be allocated. Attending is not an all-or-nothing phenomenon, but allows for gradations. Presumably, there is a default allocation among the different modalities. In this default allocation, there is enough attention allocated to vision that the visual stimuli we normally encounter are processed consciously. In contrast, in the default distribution, there is so little attention allocated to olfaction that only strong stimuli are processed consciously. However, there is *some* attention allocated to olfaction. Not actively attending to olfaction is not the same as the complete absence of attention to olfaction.

The default distribution of the attentional resource between modalities can be changed, as is shown by the ability to shift attention to modalities like olfaction or passive touch that usually receive little attention. However, the capacity to voluntarily redistribute attention is limited. It is difficult to withdraw all attention from a moving red dot in the center of the visual field. It is equally difficult to direct all attention exclusively to the feeling of one's tongue touching the roof of one's mouth for a long time. Try it! The difficulty of controlling attention is reflected by the difficulty faced when trying to master meditation techniques that require voluntarily directing and withdrawing attention. There is a strong pull back to the default distribution of attention between modalities and the conscious processes associated with it.

In summary, the amount of attention allocated to the olfactory modality and the amount of olfactory information that is processed consciously correlate closely. In the default distribution of attention between modalities, there is little attention attributed to olfaction and therefore only very strong stimuli are processed consciously. When more attention is actively allocated to olfaction, much weaker stimuli will also be consciously processed. Central to this view is that both attention and consciousness are graded rather than all-or-nothing phenomena. The close correlation between attention and conscious processing opens new experimental approaches because attention, unlike consciousness, can be systematically manipulated in psychophysical experiments. A law-like relation between attention and consciousness is waiting to be discovered.

7.3 Conclusion: The Mechanisms of Conscious Processing Are Poorly Understood

Whatever the mechanism of consciousness is, it can be described at different levels of organization. At the level of individual neurons, ethical and practical considerations limit the amount and type of data that can be collected. Only very rarely are there opportunities to record from identified neurons in the human brain during conscious processes (Fried et al. 2014). The data obtained during these opportunities is fascinating, but it is not comprehensive enough to formulate a theory about the neuronal mechanisms involved in conscious processes. The next higher level at which brain activities in humans is studied is the level of neuronal populations. Functional magnetic resonance imaging, for example, measures the activity of brain areas that contain on average 5.5 million neurons, 10^{10}synapses, and 220 km of axons (Logothetis 2008). This is a relatively good spatial resolution for non-invasive measures of brain activity. Other methods, like electroencephalography, have a lower spatial resolution, but a much higher temporal resolution than functional magnetic resonance imaging. Studying brain activities at the level of populations of several million neurons has resulted in proposals for necessary conditions for conscious processes (e.g., that they have to be in certain parts of the brain, or that they need to involve a certain degree of information integration), but not in mechanistic theories of consciousness with predictive power.

Above the level of neuronal populations is the cognitive level. I propose that in olfaction attention correlates very closely with conscious processes. However, the description of the relation between attention and consciousness that I have given here depends on endorsing one of the many different notions of attention (Taylor 2013). This reveals the weakness of any mechanistic explanation of consciousness at the cognitive level. Any disagreement with other theories about the role of attention in conscious processes is more likely to be based on unacknowledged differences in terminology than on a disagreement about facts or their interpretation.

The speculations about the mechanisms of consciousness therefore fall into two groups, neither of which is satisfying. On the one hand, there are theories about the cognitive mechanisms of consciousness. These theories often have strong predictive power; however, they do not anchor consciousness in physical activities in the brain, unless the cognitive processes are themselves defined in terms of brain activity. On the other hand, there are many theories about the neuronal mechanisms of consciousness, none of which have strong predictive power.

Note

1. Restricting the oscillations that contribute to consciousness to those with a frequency between 25 and 100 Hz (so-called gamma oscillations) also does not lead to better predictions. Gamma oscillations are common features of the information processing in brain structures that do not process information consciously, like the retina, Neuenschwander and Singer (1996). "Long-range synchronization of oscillatory light responses in the cat retina and lateral geniculate nucleus." *Nature* **379**(6567): 728–733., and the olfactory bulb. Beshel et al. (2007). "Olfactory bulb gamma oscillations are enhanced with task demands." *Journal of Neuroscience* **27**(31): 8358–8365. Gamma oscillations also persist unabated when an organism is anesthetized. Steriade et al. (1996b). "Synchronization of fast (30–40 Hz) spontaneous oscillations in intrathalamic and thalamocortical networks." *Journal of Neuroscience* **16**(8): 2788–2808. The amplitude of oscillations also does not correlate with consciousness and is sometimes higher in anesthetized animals than in awake animals. Vanderwolf (2000). "Are neocortical gamma waves related to consciousness?" *Brain Research* **855**(2): 217–224. These results indicate that no type of synchronous oscillations is limited to conscious neuronal activities. Instead, all types of oscillations are involved in all types of brain activities.

References

Atmanspacher, H. (2011). Quantum approaches to consciousness. In *The Stanford encyclopedia of philosophy*.

Beshel, J., Kopell, N., et al. (2007). Olfactory bulb gamma oscillations are enhanced with task demands. *Journal of Neuroscience, 27*(31), 8358–8365.

Bragin, A., Jando, G., et al. (1995). Gamma (40–100 Hz) oscillation in the hippocampus of the behaving rat. *Journal of Neuroscience, 15*(1), 47–60.

Brunet, N., Vinck, M., et al. (2014). Gamma or no gamma, that is the question. *Trends in Cognitive Sciences, 18*(10), 507–509.

Cameron, L. E. (2007). Measures of human olfactory perception during pregnancy. *Chemical Senses, 32*(8), 775–782.

Chalmers, D. (1995). Facing up to the problem of consciousness. *Journal of Consciousness Studies, 2*(3), 200–219.

Chalmers, D. J. (2006). Phenomenal concepts and the explanatory gap. In T. Alter & S. Walter (Eds.), *Phenomenal concepts and phenomenal knowledge: New essays on consciousness and physicalism*. Oxford: Oxford University Press.

Colgin, L. L., & Moser, E. I. (2010). Gamma oscillations in the hippocampus. *Physiology, 25*(5), 319–329.

Degel, J., & Koester, E. P. (1999). Odors: Implicit memory and performance effects. *Chemical Senses, 24*(3), 317–325.

Doty, R. L., & Cameron, E. L. (2009). Sex differences and reproductive hormone influences on human odor perception. *Physiology & Behavior, 97*(2), 213–228.

Eccleston, C. (1995). The attentional control of pain—methodological and theoretical concerns. *Pain, 63*(1), 3–10.

Fried, I., Rutishauser, U., et al. (2014). *Single neuron studies of the human brain: Probing cognition hardcover*. Cambridge: MIT Press.

Graziano, M. S. A. (2013). *Consciousness and the social brain*. New York: Oxford University Press.

Idelson, M. S., Ben-Jacob, E., et al. (2010). Innate synchronous oscillations in freely-organized small neuronal circuits. *PLoS One, 5*(12), e14443.

Koch, C., & Hepp, K. (2006). Quantum mechanics in the brain. *Nature, 440*, 611–612.

Köster, E. P., Møller, P., et al. (2014). A "Misfit" theory of spontaneous conscious odor perception (MITSCOP): Reflections on the role and function of odor memory in everyday life. *Frontiers in Psychology, 5*, 64.

Kouider, S., Gardelle, V., et al. (2010). How rich is consciousness? *Trends in Cognitive Sciences, 14*, 301–307.
Krauel, K., Pause, B. M., et al. (1998). Attentional modulation of central odor processing. *Chemical Senses, 23*(4), 423–432.
Levine, J. (1983). Materialism and qualia: The explanatory gap. *Pacific Philosophical Quarterly, 64*, 345–361.
Logothetis, N. K. (2008). What we can do and what we cannot do with fMRI. *Nature, 453*(7197), 869–878.
Lorig, T. (1992). Cognitive and 'non-cognitive' effects of odor exposure: Electrophysiological and behavioral evidence. In S. Van Toller & G. Dodd (Eds.), *The psychology and biology of perfume* (pp. 161–173). Amsterdam: Elsevier.
Mole, C. (2008). Attention and consciousness. *Journal of Consciousness Studies, 15*(4), 86–104.
Muramoto, K., Ichikawa, M., et al. (1993). Frequency of synchronous oscillations of neuronal activity increases during development and is correlated to the number of synapses in cultured cortical neuron networks. *Neuroscience Letters, 163*(2), 163–165.
Neuenschwander, S., & Singer, W. (1996). Long-range synchronization of oscillatory light responses in the cat retina and lateral geniculate nucleus. *Nature, 379*(6567), 728–733.
Norman, L. J., Heywood, C. A., et al. (2013). Object-based attention without awareness. *Psychological Science, 24*(6), 836–843.
O'Brien, G., & Opie, J. (1999). A connectionist theory of phenomenal experience. *Behavioral and Brain Sciences, 22*, 127–148.
Overgaard, M., Fehl, K., et al. (2008). Seeing without seeing? Degraded conscious vision in a blindsight patient. *PLoS One, 3*, 1–4.
Pham, J., Pakdaman, K., et al. (1998). Noise-induced coherent oscillations in randomly connected neural networks. *Physical Review E, 58*(3), 3610–3622.
Prinz, J. J. (2012). *The conscious brain: How attention engenders experience.* Oxford: Oxford University Press.
Sela, L., & Sobel, N. (2010). Human olfaction: A constant state of change-blindness. *Experimental Brain Research, 205*(1), 13–29.
Seth, A. K., Izhikevich, E., et al. (2006). Theories and measures of consciousness: An extended framework. *Proceedings of the National Academy of Sciences, 103*(28), 10799–10804.
Singer, W. (2011). Consciousness and neuronal synchronization. In S. Laureys & G. Tononi (Eds.), *The neurology of consciousness* (pp. 43–52). New York: Academic.

Spence, C., & Driver, J. (1997). On measuring selective attention to an expected sensory modality. *Perception & Psychophysics, 59*, 389–403.
Spence, C., McGlone, F., et al. (2001). Attention to olfaction – A psychophysical investigation. *Experimental Brain Research, 138*(4), 432–437.
Steriade, M., Amzica, F., et al. (1996a). Synchronization of fast (30–40 Hz) spontaneous cortical rhythms during brain activation. *Journal of Neuroscience, 16*(1), 392–417.
Steriade, M., Contreras, D., et al. (1996b). Synchronization of fast (30–40 Hz) spontaneous oscillations in intrathalamic and thalamocortical networks. *Journal of Neuroscience, 16*(8), 2788–2808.
Taylor, J. (2013). Is attention necessary and sufficient for phenomenal consciousness? *Journal of Consciousness Studies, 20*(11–12), 173–194.
Tononi, G. (2008). Consciousness as integrated information: A provisional manifesto. *Biological Bulletin, 215*, 216–242.
Uhlhaas, P. J., Pipa, G., et al. (2009). Neural synchrony in cortical networks: History, concept and current status. *Frontiers in Integrative Neuroscience, 3*, 17.
van Boxtel, J. J. A., Tsuchiya, N., et al. (2010). Consciousness and attention: On sufficiency and necessity. *Frontiers in Psychology, 1*, 217.
Vanderwolf, C. H. (2000). Are neocortical gamma waves related to consciousness? *Brain Research, 855*(2), 217–224.
Zelano, C., Bensafi, M., et al. (2005). Attentional modulation in human primary olfactory cortex. *Nature Neuroscience, 8*(1), 114–120.

8

Function of Conscious Brain Activities

Many researchers have wondered what, if anything, the function of consciousness is (Seth 2009; Van Gulick 2011). I will not contribute to this discussion here and instead speculate about the function of conscious brain activities. This may appear to be an unnecessarily fine distinction, but in Sect. 8.1. I will show that this distinction is necessary to bring the study of function in consciousness research in line with the study of function in other areas of biology. I will then discuss the notion of function in biology with a special emphasis on the fact that a lack of alternatives is not part of the notion of function in biology or anywhere else. At the end of Sect. 8.1., I will suggest contrastive analysis as a method to determine the evolutionary function of conscious brain activities. After these methodological considerations, I will apply contrastive analysis to identify the evolutionary function of conscious brain activities in Sect. 8.2. In Chap. 4, I have argued that it is the function of perception to guide behaviors. The function of conscious perception therefore has to be part of guiding behavior. A comparison between situations in which olfactory information is processed consciously and situations in which it is not processed consciously reveals that the function of conscious processes is to guide behaviors in situations with many behavioral options.

A. Keller, *Philosophy of Olfactory Perception*,
DOI 10.1007/978-3-319-33645-9_8

8.1 Determining the Function of Conscious Brain Activities

Many disagreements about function in consciousness research are not disagreements about the data or its interpretation, but disagreements about terminology and methodology. In many other fields of biological research, functions have been identified successfully. These fields provide tested and widely accepted terminologies and methodologies. My strategy is to use this established framework of the analysis of functions in biology and apply it to conscious brain activities. Questions about functions in consciousness research should be, at least as a first attempt, approached in the same way in which these questions have been addressed in hematology or ornithology.

Function of Consciousness or Function of Conscious Brain Activities

I have said above that I will not discuss the function of consciousness, but instead the function of conscious brain activities. The difference between these two projects is that consciousness is a property of some brain activities, whereas conscious brain activities are a process. Like in other fields of biology, in consciousness research, one cannot consider the function of an isolated property of a process. For mechanisms, processes, and structures, the starting point to investigate their function is to imagine, or empirically test, the consequences of the removal of the mechanism, process, or structure. To find out what the function of birds' wings is one has to find out what birds cannot do without their wings. For the function of a given property, this approach is often not possible because the property of interest often cannot be removed from the mechanism, process, or structure without affecting other properties.

That trying to identify the function of an isolated property is often a non-starter can be illustrated with the example of red blood cells' property of being red. There are red and white blood cells. Redness is a property of red blood cells. One can ask two different questions about function in relation to red blood cells. One can ask either what the function of red blood cells is or what the function of the "redness" of red blood cells is. The same is true for consciousness. One can ask either what the function of conscious brain activities is, or what the function of the consciousness of these brain activities is. If one would ask a hematologist what the function of red blood cells is, she would respond that the function of red blood cells is distributing oxygen from the lungs to other body tissues. If one would ask her what the function of the redness of red blood cells is, she would presumably explain that red blood cells are red because they are rich in iron-containing hemoglobin, which is red. This is an explanation, but it does not assign a function to the property "redness".

The hematologist fails to assign a function to the "redness" of red blood cells because a situation in which the color of the cells changes whereas all other properties remain the same is inconceivable. As long as the cells contain hemoglobin, they are red. It is equally impossible for a brain activity to change from being conscious to non-conscious while all its other properties remain unchanged. In the philosophical literature, the scenario of changing conscious brain activities into non-conscious brain activities without changing anything else about the brain activities is illustrated by the "philosophical zombie". A philosophical zombie is an exact physical copy of a human, but none of its brain activities is conscious. To see why a philosophical zombie is inconceivable, consider its hematological equivalent, the hematological zombie. A hematological zombie is a creature with blood cells that are exact physical copies of red blood cells, but they are not red. Hematological zombies are neither possible nor conceivable. The redness of hemoglobin is a consequence of how it reflects light. How a molecule reflects light depends on its structure. The only way to change the redness of a molecule is therefore to change its structure, which makes it physically different from hemoglobin. The redness of the cells cannot be changed without changing other things about them. Philosophical zombies are inconceivable for analogous reasons. Changing a brain activity from conscious to non-conscious involves changing other properties of the brain activity as well. Only Cartesian dualists can remove the property "being conscious" from brain activities without changing any of their other properties.[1]

Hematology is a mature research field. The struggles of hematologists to identify the function of the redness of red blood cells are not due to some unresolved mystery about the biology of blood cells. Instead, the question what the function of the redness of red blood cells is, is a bad question because it is neither conceptually nor experimentally possible to isolate the redness and consider or study its function in isolation. There are other similarly bad questions. What is the function of a bacterium being alive? What is the function of the weight of our liver? What is the function of a brain activity being conscious? What is the function of stomach acid dissolving metal? What is the function of DNA replication being semiconservative? Each of these questions are difficult to answer not because its topic is poorly understood, but because they are questions about the function of properties of mechanisms, processes, or structures that cannot be considered independently of other properties.

"What is the function of the redness of red blood cells?" is a question that hematologists will struggle to answer. Instead of giving an answer, they are likely to explain why the structure of the red blood cells, which is dictated by their function, results in them being red. They will not be able to give a different answer regardless of how dramatically our understanding of blood cells will increase. In contrast to these difficulties associated with discussions of the function of a property, "what is the function of red blood cells?" is a question that every hematologist is ready to answer. Finding the answer to this question was a significant discovery in hematology. Consciousness research should follow this and other success stories and search for the function of entities or mechanisms rather than the function of isolated properties. The interesting and answerable question about function and consciousness is "What is the function of conscious brain activities?"

The Notion of Function in Biology

"What is the function of conscious brain activities?" is an ambiguous question because there are diverse philosophical theories about what "functions" in biology are (for a collection of essays on the topic, see Buller 1999). Functions can be either teleological functions, or causal (or systemic) functions. The teleological function of a biological structure or mechanism is what it was selected for (Millikan 1984). The causal function of a mechanism is the role of a structure or mechanism within a complex system (Cummins 1975). I will consider here the teleological function of conscious processes.

The teleological function of something is the same as its adaptive value or the reason why it was selected through natural selection. A complication when considering teleological function is that one has to distinguish between the current utility and the reason for origin. The reason why there is selective pressure on ostriches to maintain their wings is that they use their wings to steer when they are running and for courtship and dominance displays. This is the reason why the ostrich wings are not going away. However, the reason why ostrich wings came into being in the first place is that some ancestors of ostriches used them for flight. The modern history approach to functions considers the recent past and the explanation why a structure of mechanism was maintained over the recent past (Godfrey-Smith 1994). According to the modern history theory of function, the function of the ostriches' wings is steering during running and courtship. I will deviate from this approach and instead speculate about the reason why conscious processes emerged, rather than why they have been maintained through recent evolutionary history. The account here is meant as a speculation about how conscious perception was used when it first appeared.

A Lack of Alternatives Is Not Part of the Notion of Function

The bar for what qualifies as a function of conscious brain activities is often set very high. It is frequently expected that the function of conscious brain activities is something that non-conscious brain activities cannot accomplish. Underlying this expectation is a simplified model of evolution through natural selection. The first creature capable of conscious processing supposedly must have had the capacity to do things that its non-conscious ancestors and competitors were unable to do. Being able to perform this novel function gave the conscious creature an adaptive advantage that resulted in the spread of conscious processes through the population. While this is a possible scenario, it is not the only scenario and neither the lack of non-conscious alternative realizations nor having a novel function is a necessary criterion for something to be the function of conscious brain activities. Instead, to say of X that it is the function of conscious processes means only that conscious processes are there because they do X, and that X is a consequence of conscious processes (Wright 1976).

The inference from the existence of alternative mechanisms to the absence of function is never encountered outside of the field of consciousness research. One can sit on rocks, benches, and toilets, but this does not change that it is the function of chairs to provide a surface to sit on. The lack of alternatives is also not considered a requirement for the more technical notion of evolutionary function in biology. The function of fishes' fins is swimming although many mammals, birds, insects, amphibians, jellyfish, and other creatures swim without fins. As has been pointed out previously (Dretske 1997), what makes something the function of a biological mechanism is not the absence of alternatives.

The idea of alternative mechanisms to conscious processes is in philosophy of mind illustrated by the behavioral zombie, which is a variant of the philosophical zombie. Behavioral zombies are creatures that behave exactly like humans, but their brains are wired differently and none of their brain activities is accompanied by consciousness. If the lack of alternatives were considered part of the definition of "function", then the possibility of behavioral zombies would show that conscious brain activities have no function. The hematological analog to the behavioral zombie is a creature in which oxygen is distributed from the lungs just like in normal humans, but without the involvement of red blood cells. One can easily imagine such a creature. The red oxygen-binding hemoglobin could be replaced with green or blue oxygen-binding proteins that are found in worms and crustaceans. Oxygen could be distributed in our body in the absence of red blood cells. However, for an investigation into the function of red blood cells in humans, the possibility of alternative realizability of the red blood cells' function is unimportant. That other cells could distribute oxygen does not mean that it is not the function of red blood cells to distribute oxygen.

If the lack of alternatives would be a requirement for something to be a function then neither red blood cells nor conscious brain activities would have a function. In fact, applying this criterion would lead to the conclusion that nothing in biology has a function. Different animal species often evolved in diverse ways to solve the same problem. The variety of molecular, physiological, and behavioral mechanisms discovered by comparative biologists is astonishing. Only the very basic biological mechanisms and structures are conserved between all living things. However, even at this level, things *could* be implemented differently. Even though a DNA genome is widely shared among living things, many other molecules, like the similar RNA, could be used to store inheritable information. No geneticist concludes from this insight that genomes made of DNA have no function.

Why did we evolve to have conscious brain activities when we could have done everything we are doing by processing information non-consciously? Several evolutionary scenarios could explain this fact. Consider the analogy between evolution through natural selection and the competition between goods in an idealized marketplace. A new computer model may replace the older models because it can do something new. Maybe the new model is the first to have a three-dimensional display. If consumers like three-dimensional displays, the new model will replace the older models. However, if the new model is ten times more expensive than the older models, consumers may be reluctant to pay that much more and the new model will fail to replace the older models. Alternatively, a new computer model may replace the older models because it is cheaper, although it can do nothing the older models cannot also do. Even a new model that has fewer functions than the older models may replace the older models, if it is much cheaper. This analogy shows that novelty is not intrinsically adaptive whereas efficiency is.

When it comes to brain activities, efficiency is achieved by keeping the brain small and its energy expenditure low. Imagine that a behavioral zombie that behaves like a human but processes all information non-consciously is possible, but only with a brain ten times larger than the human brain. Such a large brain would pose problems for stabilizing the head and for movement. It would also be expensive to build and maintain. Therefore, even if the problems that are solved in our brains by conscious processes could be solved non-consciously, but only in a larger brain, there would have been strong adaptive pressure to process information consciously.

Conscious information processing could even have evolved in a situation in which non-conscious information processing is as efficient as conscious information processing. It is possible that historical contingencies are the reason why there are conscious brain activities. In this scenario, the things that are performed by conscious brain activities could be performed with equal efficiency by non-conscious activities. However, the conscious solution happened to be "found" first through the process of random generation of variability. Then, after this solution has been established, the alternatives have a much higher entry barrier. Think of it as a complex computer program written in a suboptimal programming language. Even when better languages are developed later, it is easier to continue updating the program in the suboptimal language than to rewrite it in its entirety in the new language.

All these considerations about possible scenarios of the evolution of conscious brain activities are pure speculations not based on any evidence. I do not intend to trace back the actual evolutionary history of conscious processes. Instead, these hypothetical yet possible evolutionary histories illustrate that the lack of alternatives is not a necessary requirement for something to be the function of conscious brain activities. No considerations about possibilities, probabilities, or efficiency play any role in determining what the function of conscious processes is. Instead, the function of conscious processes is determined by the reasons for which information was initially processed consciously.

Contrastive Analysis

Since the existence or lack of alternatives is irrelevant for uncovering the function of conscious brain processes, the large research project of trying to identify behavioral tasks that can only be accomplished using conscious processes is not providing any evidence about the function of the conscious processes. Instead, it is necessary to determine the adaptive advantage that the first organism that ever processed information consciously had. Obviously, this cannot be done based on direct observations. Brains, minds, and behaviors do not fossilize. This is, however, not an unusual situation in biology. The adaptive advantage that was conveyed by the first wing or the first fin can also not be observed directly. The indirect strategy for identifying the evolutionary function of such traits is contrastive analysis. Contrastive analysis compares situations in which the mechanism under study is employed with situations in which alternative mechanisms are employed. For fish's fins, this methodology would result in identifying aquatic locomotion as the fins' evolutionary function. Contrastive analysis requires generalizations over many cases. Evolution is an ongoing process and the correlation between traits and functions cannot be expected to be perfect. The fact that some animals without fins are capable of aquatic locomotion does not mean that aquatic locomotion is not the function of fins. Furthermore, aquatic locomotion is the evolutionary function of fins even though a contrastive analysis is likely to uncover that sometimes fins are not used for aquatic locomotion but for walking over land, courtship displays, or temperature regulation. It is very common for structures or mechanisms that evolved for one function to be further adapted for additional functions. The goal of contrastive analysis is to analyze current uses to identify the phylogenetically earliest function of a structure or mechanism. If the evidence shows that the first animals with fins used them for aquatic locomotion, then aquatic locomotion is the evolutionary function of fins. To identify the function of conscious processes, conscious and non-conscious processing of the same stimulus has to be compared.

8.2 Function of Conscious Brain Activities in Olfaction

As I have shown in Chap. 4, the function of perception, including olfactory perception, is to guide behaviors. In part III, I have then argued that different sensory modalities are differentially connected to cognitive processes. This differential connectivity reflects the fact that perception in different modalities has different functions. Perception always has the function of guiding behaviors, but which behaviors are guided differs between modalities. Vision is the Swiss Army knife of the sensory modalities and it is used for guiding a wide variety of behaviors.

Other modalities are more specialized. The function of taste and somatosensation are rather different. Gustatory perception guides food intake behavior, whereas somatosensory perception has the function of guiding stabilizing body movements. Olfaction, as discussed in Chap. 5, is predominantly an evaluative sense in humans. It has been suggested that the most important things that humans evaluate using their sense of smell is food (Shepherd 2011).[2] Other humans, potential dwellings, and other things are also evaluated olfactorily (Stevenson 2009).[3] The evolutionary function of an evaluative sense is the guiding of decision behaviors.[4] That olfaction is primarily an evaluative sense is true for humans. However, in many other mammals, the most prominent odor-guided behavior is navigation (Jacobs 2012). Although odor-guided navigation plays no important role in humans, as I have discussed in Sect. 3.1., there are some examples of it. Infants, for example, use olfactory cues to orient toward their mother's breast (Varendi et al. 1994; Varendi and Porter 2001). Under experimental condition, humans are also surprisingly good at following an odor trail (Porter et al. 2007).

The function of conscious processing of olfactory information has to contribute to the overall function of olfactory perception, which is to guide decision behaviors. To identify the function of conscious processing of olfactory information, it is necessary to compare cases in which olfactory information is processed consciously to cases in which the same information is not processed consciously. It is important that the stimulus is the same in the compared cases. Conscious and non-conscious processes in the visual system are often compared by contrasting the processing of two different stimuli, for example, a short visual display (that is not processed consciously) and a long visual display (that is processed consciously). These types of experiments are not very useful for elucidating the function of conscious processes because any difference in how the two different stimuli affect behavior may be caused either by the difference in the stimuli or by the fact that one of the stimuli is processed consciously whereas the other is not. Comparative analysis of the processing of *identical* physical stimuli is therefore preferable (for a more detailed discussion of this point, see Kim and Blake 2005).

Olfaction is an ideal system for comparing different levels of consciousness of the processing of physically identical stimuli because both conscious and non-conscious perception of smells is common in everyday life. Our olfactory sensory neurons are frequently activated by odor molecules and usually this information is processed non-consciously. Only sometimes, when it is attended to, is the information processed consciously. In olfaction, awareness of the stimulus is the exception rather than the rule (Köster 2002).

Odor-Guided Behaviors

Often, making decisions based on olfactory information does not require conscious processing. This is reflected by olfactory metaphors for situations in which we make a decision without having conscious awareness of our reasons. We say that we "smell a rat", or that "something smells fishy". Other things may pass the "smell test". That olfactory evaluation does not always require conscious processes has also been demonstrated empirically. Social preferences, for example, have been shown to be influenced by odors that were not consciously processed by the subjects (Li et al. 2007). Similarly, there is a non-conscious effect of odors on judgments of participants posing as job candidates (Cowley et al. 1977). Like evaluation of other people, evaluation of food often does not require conscious neuronal activities. For example, sucrose solution is evaluated to be sweeter when a consciously undetected small amount of the pineapple odor ethyl butyrate is added (Labbe et al. 2006). Similarly, odors at concentrations that are too low to be consciously processed can change the perceived odor quality when added to a mixture (Guadagni et al. 1963; Ito and Kubota 2005). In all these cases, consciousness is not required for evaluation and for the decision behavior. However, there are also tasks in which conscious information processing is required to make a decision. When the decision that has to be made is to either swallow or spit out a sip of wine, conscious processing is not required. However, when the task is to write a review of the wine's flavor, and the decision that has to be made is what words to use to describe the wine, it is necessary to process the sensory information consciously.

Humans use their sense of smell predominantly for evaluation, but they are also capable of odor-guided navigation. The only strategy available to locate the source of the gas leak in a building is through serial sampling and comparisons (Unlike other species, humans do not have the capacity for directional smelling by comparing the olfactory input of the two nostrils (Radil and Wysocki 1998; Frasnelli et al. 2008; Kleemann et al. 2009)). To locate the gas leak, one has to sample the air by sniffing while walking from room to room. Through intensity comparisons, the location of the gas leak can be identified (Richardson 2011). Throughout the entire process, olfactory information is processed consciously and compared to stored conscious percepts of the smell in the other rooms. It seems unlikely that this task could be accomplished without conscious information processing. On the other hand, there is evidence that odor-dependent place preferences can be mediated without conscious processing of the sensory information. It has been shown that people chose chairs in a dentist's waiting room depending on the odor the chairs were perfumed with (Kirk-Smith and Booth 1980; Pause 2004). In this study, subjects were not aware of the odor. In another study, perfuming a small pizzeria in the Brittany region of France with lavender increased the time patrons spent in the restaurant as well as the amount of money they spent (Guéguen and Petr 2006). Many studies of the effect of ambient scents on behaviors do not control for all potential biases (Teller and Dennis 2012) and subject numbers are usually low. Replications are rare. Another complication of this type of studies is that it is very difficult to demonstrate that olfactory processing was *completely* non-conscious. Just because there is no memory of a lavender odor after dinner does not establish that the odor was at no point during the dinner consciously processed.[5] Each individual study has, therefore, to be interpreted with care. However, I think that taken together there is good evidence that we rather spend time in a pleasantly scented area than in an unpleasantly scented area, and that this preference can be mediated through non-conscious processing.

Which Situations Require Conscious Activity in the Olfactory System?

The examples from odor-based evaluation and odor-guided navigation show that in some situations odor-guided decision-making requires conscious processes while in other situations it does not. The salient difference between situations in which information is processed with a high level of consciousness and situations in which conscious processing is not required is the number of behavioral options between which the organism has to choose (Keller 2014). When there are only two behavioral options, spitting or swallowing, then information about the wine does not have to be processed consciously. In situations in which there are a large number of options, like when a review has to be written, then conscious processes are required. Similarly, in the case of having a place preference based on an odor, there are only two options: stay/go. However, if the location of the odor source has to be identified, then there are as many options as there are paths in two-dimensional space. In these examples, the number of behavioral options increases dramatically because of the large number of possible combinations of steps or words that make up paths or reviews. The task of writing a review consists of deciding between the astronomically large number of possible combinations of words. Similarly, every navigation in space is a combination of many stay/go/turn decisions. Behavioral decisions, in which such combinations are required, require conscious processing. Verbal communication and goal-directed navigation in physical space are combinatorial tasks with a very large number of options, which is why they require conscious information processing.[6]

The proposal that information is processed consciously when an organism is faced with many behavioral options explains why olfactory information, compared to visual information, is often processed non-consciously. Behaviors that are visually guided are usually more complex than those that are odor-guided. Vision is the dominant sense in humans because it represents physical space more accurately than the other senses. Behaviors that depend on precise movements in physical space, like manipulation of objects and tool use, usually require choosing between a large number of behaviors and the visual information that guides these behaviors is therefore most efficiently processed consciously. In contrast, as pointed out in Sect. 5.2., olfaction mostly guides evaluative behaviors, which are usually associated with binary decisions like stay/go, spit/swallow, inhale/hold your breath, or approach/avoid.

Counterexamples

Counterexamples to the proposal that it is the function of conscious processes to mediate behavioral decisions when there are many different options are easily found. Sometimes, when taking a nice fragrant bubble bath after a long day, we experience the smell of the bubble bath although we are not about to make any decisions. At other times, we consciously experience the pain from touching a hot plate on the oven although the behavioral decision we have to make is simple. We either retract the hand or leave it on the hot oven plate. These examples show that information is processed consciously in many situations in which no complex decisions are made. However, the apparent counterexamples are not in conflict with the proposal that it is the evolutionary function of conscious olfactory processing to guide behaviors in situations with many different behavioral options.

To illustrate this, let us return to the example of how ornithologists arrived at the conclusion that the function of bird wings is flight. Wings, like conscious information processing, are a tool that evolved for a specific purpose. In the case of wings, that purpose is flying. After wings became available, it was possible to use them for things other than flying, too. Pelicans use their wings to beat on the water surface to drive their prey into the shallows. Males of some jacana species carry their chicks under their wings. Male birds of paradise use their wings during courtship displays, and cranes use their wings to shade the water surface to better see their prey swimming underneath (Gazzaniga et al. 2009, p. 651). Ostriches use their wings to steer when they are running and for courtship and dominance displays. In penguins, wings are used for swimming. According to the modern history approach to biological functions, this means that the functions of wings are different in ostriches, penguins, and sparrows. It is difficult to see how the modern history approach could answer the question about the function of wings in general, since the reasons why wings are maintained differ between bird species. This is why I approached the question of functions by considering why wings first evolved. Considering how ubiquitously wings are used for flying, contrastive analysis suggests that wings first evolved for the purpose of flying.

Consciously processing the smell of the lavender-lemon bubble bath is the equivalent of a bird walking around and flapping its wings. It is not clear what the function of the wing flapping could be. It may have a function, for example, in temperature regulation. But it also may have no function; it may be that the bird flaps its wings while walking around as part of learning how to control them in situations in which they do have a function. Or maybe the bird flaps the wings because they are itchy, or the bird has muscle spasms. On many occasions, birds can be observed moving their wings although they are not flying. In the same manner, we often use conscious information processing although we are not making any decisions.

Consciously processing the pain from the hot oven plate is the equivalent of a bird using its wings to defend itself against an attacking cat. When a bird is attacked, it will use all available means to defend itself. Whether a bird's peak, claws, or wings evolved for self-defense is irrelevant for the bird in such a situation. In analogy, a very strong heat stimulus is an indication of a possibly life-threatening situation. In such a situation, all available tools, including conscious information processing, are used to find the appropriate behavioral response.

That in some situations information is processed consciously although no decisions between many behavioral options are made does not show that it is not the evolutionary function of conscious processing to make such a decision. Contrastive analysis to determine the evolutionary function of a trait does not require that the evolutionary function is the only function ever observed. Every counterexample can be addressed within this framework. My speculations about why birds flap their wings and people in bathtubs perceive the smell of the bubble bath consciously are probably wrong. However, the important point is that any counterexample can be addressed within the framework of contrastive analysis. The method is one of weighing evidence.

Relevant Behavioral Options

One potential objection to the proposal that the function of conscious brain activities is to decide between behavioral options when there are many options is that the total number of behavioral options does not change dramatically between different situations. The above example of a situation with few behavioral options was to drink wine for sustenance. I said that in this situation, there are only two behavioral options, spitting or swallowing. In contrast, when the flavor of wine is perceived with the goal of writing a review about it, there are as many different behavioral options as there are possible wine reviews. This is not strictly true. Instead, for whatever reason we drink wine, we always have the same number of behavioral options. We can always spit out the wine or write a review about it (or compose a review in our mind). We also have many other behavioral options, like jumping up and down or making monkey noises. The number of possible behavioral options is only limited by physical constraints.

The answer to this objection is to refine the proposal from *possible* behavioral options to *goal-relevant* behavioral options. Writing a review or making monkey noises are not goal-relevant behavioral options when I drink wine with the goal to quench my thirst. Everyday experience outside of olfactory perception confirms that we perceive things consciously in situations with many goal-relevant behavioral options. When we drive a familiar route in low traffic, little sensory information is processed consciously. However, if suddenly a deer jumps in front of the car, information has to be processed consciously, because the deer makes it necessary to consider a wide variety of possible responses to avoid a collision. The total number of behavioral options has not changed. We still sit in the same car and it is possible to turn the steering wheel in any direction, to hit the brakes, to switch on the radio, etc. However, the number of relevant behavioral options has changed dramatically.

That the number of *relevant* behavioral options determines whether it is necessary to process information consciously or not also explains why there are situations in which the same information during the same task has to be consciously processed by some, but not by others. During skill acquisition information has to be processed consciously whereas during skill retrieval the same information can efficiently be processed non-consciously (Schneider et al. 1994; Floyer-Lea and Matthews 2004). As someone learns to play a new song on the guitar, they have to process their finger positions and movements consciously. However, as they become more familiar with the song, the finger movements can increasingly be guided by non-conscious information processing. Because skill acquisition is a gradual process, it reveals nicely the gradual nature of consciousness. With practice, playing the song requires continuously less conscious information processing. The reason for this change in how the information is processed is that familiarization with the song decreases the number of relevant behavioral options. When a song is played from sheets for the first time, at every point during the song, a very large number of combinations of notes and therefore finger movements may follow. Once the song is familiar, only one sequence of finger movements is pertinent. The number of relevant behavioral options is reduced to one, which abolishes the need for conscious processing of information.[7]

8.3 Conclusion: The Function of Conscious Processes Is to Facilitate Decision Making in Situations with Many Behavioral Options

I have identified decision making in situations with a large number of behavioral options as the function of conscious brain processes. I do not anticipate readers that are familiar with works on the function of consciousness to be impressed with this result because it is mainly based on changing the question and the requirements for a satisfying answer. The purpose for these changes was to bring questions about function in consciousness science in line with questions about function in other fields of biology. One consequence of these changes was that instead of discussing the function of the property of being conscious, I discussed the function of conscious processes. This change is analogous to the switch from studying the function of red blood cells' property of being red to studying the function of red blood cells.

Another consequence of the methodological change that I propose is that, according to me, for something to be the function of conscious processes, it is not required that there are no possible other processes that could also have that function. Furthermore, counterexamples are not arguments against something being the function of a process.

When the function question in consciousness research is demystified and brought in line with the study of functions in other areas of biology, simple observation in the processing of olfactory information shows that whether the information is processed consciously depends not on the stimulus (although the stimulus must be strong enough). The same stimulus is processed consciously in some and non-consciously in other situations. The salient difference between situations that require conscious processes and those that do not is that in the situations that require conscious information processing, there are more relevant behavioral options than in the situations that do not require conscious processing.

Notes

1. For Cartesian dualists, the question of the evolutionary function of consciousness does not arise because for them consciousness is uncoupled from the physical world, including the genome, which is the substrate of evolution through natural selection.
2. The main function of olfaction would therefore be the same as the function of gustation: to guide feeding behavior as a component of flavor perception.
3. There is a tendency to think of our ability to smell as not being functionally important at all. The olfactory system is sometimes treated like a vestigial organ, the sensory equivalent of the appendix. Aristotle, in *De Anima*, writes that, " our sense of smell is not accurate but worse than many animals'. For man smells poorly". Charles Darwin wrote about the human sense of smell that it "is of extremely slight service, if any, even to savages, in whom it is generally more highly developed than in the civilised races. It does not warn them of danger, nor guide them to their food; nor does it prevent the Esquimaux from sleeping in the most fetid atmosphere, nor many savages from eating half-putrid meat" Darwin (1871). *The Descent of Man, and Selection in Relation to Sex*, John Murray. The low opinion of olfaction's usefulness is partially due to misattributing the contribution of smell to food intake behavior to taste and partially due to the fact that olfaction often functions outside of awareness.

4. Note that, following the discussion of the function of perception in Chap. 4, evaluation cannot be an evolutionary function. To evaluate something brings no adaptive advantage. Only when there are adaptive behaviors guided by the evaluative perception will the chances of survival of the perceiver increase.
5. The level of consciousness can change gradually and even in laboratory experiments that have been designed to include objective evidence for the complete absence of consciousness in perception there is evidence that there is weak conscious experience. Sandberg et al. (2014). "Evidence of weak conscious experiences in the exclusion task." *Frontiers in Psychology* **5**(1080).
6. My proposal is not that conscious processing is involved in adjudicating goal conflicts, which has been suggested by others. Morsella et al. (2015). "Homing in on consciousness in the nervous system: An action-based synthesis." *Behavioral and Brain Sciences*. The complexity that requires conscious processes to be tackled is not a consequence of conflicting goals, but of the difficulty to decide which of the large number of possible behavioral sequences will reach the goal.
7. Related to the notion of relevant behavioral options is the notion of flexible action. An increase in relevant behavioral options increases behavioral flexibility. Behavioral flexibility has been discussed under diverse names and with different degrees of commitment to specific mechanistic theories of consciousness as a function of consciousness (for a review, see Seth2009). Functions of consciousness. *Elsevier Encyclopedia of Consciousness*. W. P. Banks, Academic Press: 279–293.; for interesting examples, see Ramachandran and Hirstein (1997). "Three laws of qualia: Clues from neurology about the biological functions of consciousness and qualia." *Journal of Consciousness Studies* **6**: 15–41, Morsella (2005). "The function of phenomenal states: Supramodular interaction theory." *Psychological Review* **112**(4): 1000–1021.).

References

Buller, D. J. (Ed.). (1999). *Function, selection, and design*. Albany: SUNY Press.

Cowley, J. J., Johnson, A. L., et al. (1977). The effect of two odorous compounds on performance in an assessment-of-people test. *Psychoneuroendocrinology, 2*(2), 159–172.

Cummins, R. (1975). Functional analysis. *The Journal of Philosophy, 72*, 741–765.

Darwin, C. R. (1871). *The descent of man, and selection in relation to sex*. London: John Murray.

Dretske, F. I. (1997). What good is consciousness? *Canadian Journal of Philosophy, 27*(1), 1–15.

Floyer-Lea, A., & Matthews, P. M. (2004). Changing brain networks for visuomotor control with increased movement automaticity. *Journal of Neurophysiology, 92*, 2405–2412.

Frasnelli, J., Charbonneau, G., et al. (2008). Odor localization and sniffing. *Chemical Senses, 34*(2), 139–144.

Gazzaniga, M. S., Ivry, R. B., et al. (2009). *Cognitive neuroscience: The biology of the mind*. New York: W. W. Norton & Company.

Godfrey-Smith, P. (1994). A modern history theory of functions. *Noûs, 28*, 344–362.

Guadagni, D. G., Buttery, R. G., et al. (1963). Additive effect of subthreshold concentrations of some organic compounds associated with food aromas. *Nature, 200*, 1288–1289.

Guéguen, N., & Petr, C. (2006). Odors and consumer behavior in a restaurant. *International Journal of Hospitality Management, 25*(2), 335–339.

Ito, Y., & Kubota, K. (2005). Sensory evaluation of the synergism among odorants present in concentrations below their odor threshold in a Chinese jasmine green tea infusion. *Molecular Nutrition & Food Research, 49*(1), 61–68.

Jacobs, L. F. (2012). From chemotaxis to the cognitive map: The function of olfaction. *Proceedings of the National Academy of Sciences, 109*(Supplement_1), 10693–10700.

Keller, A. (2014). The evolutionary function of conscious information processing is revealed by its task-dependency in the olfactory system. *Frontiers in Psychology, 5*, 62.

Kim, C.-Y., & Blake, R. (2005). Psychophysical magic: Rendering the visible 'invisible'. *Trends in Cognitive Sciences, 9*(8), 381–388.

Kirk-Smith, M. D., & Booth, D. A. (1980). *Effect of androstenone on choice of location in others' presence. Olfaction and taste* (Vol. 7, pp. 397–400). London: IRL Press Limited.

Kleemann, A. M., Albrecht, J., et al. (2009). Trigeminal perception is necessary to localize odors. *Physiology & Behavior, 97*(3–4), 401–405.

Köster, E. P. (2002). The specific characteristics of the sense of smell. In C. Rouby, B. Schaal, D. Dubois, R. Gervais, & A. Holley (Eds.), *Olfaction, taste, and cognition* (pp. 27–44). Cambridge: Cambridge University Press.

Labbe, D., Rytz, A., et al. (2006). Subthreshold olfactory stimulation can enhance sweetness. *Chemical Senses, 32*(3), 205–214.
Li, W., Moallem, I., et al. (2007). Subliminal smells can guide social preferences. *Psychological Science, 18*(12), 1044–1049.
Millikan, R. (1984). *Language, thought, and other biological categories*. Cambridge: MIT Press.
Morsella, E. (2005). The function of phenomenal states: Supramodular interaction theory. *Psychological Review, 112*(4), 1000–1021.
Morsella, E., & Godwin, C. A. et al. (2015). Homing in on consciousness in the nervous system: An action-based synthesis. *Behavioral and Brain Sciences, 22*, 1–106.
Pause, B. (2004). Are androgen steroids acting as pheromones in humans? *Physiology & Behavior, 83*(1), 21–29.
Porter, J., Craven, B., et al. (2007). Mechanisms of scent-tracking in humans. *Nature Neuroscience, 10*(1), 27–29.
Radil, T., & Wysocki, C. J. (1998). Spatiotemporal masking in pure olfaction. *Annals of the New York Academy of Sciences, 855*, 641–644.
Ramachandran, V. S., & Hirstein, W. (1997). Three laws of qualia: Clues from neurology about the biological functions of consciousness and qualia. *Journal of Consciousness Studies, 6*, 15–41.
Richardson, L. (2011). Sniffing and smelling. *Philosophical Studies, 162*(2), 401–419.
Sandberg, K., Del Pin, S. H., et al. (2014). Evidence of weak conscious experiences in the exclusion task. *Frontiers in Psychology, 5*, 1080.
Schneider, W., Pimm-Smith, M., et al. (1994). Neurobiology of attention and automaticity. *Current Opinion in Neurobiology, 4*, 177–182.
Seth, A. K. (2009). Functions of consciousness. In W. P. Banks (Ed.), *Elsevier encyclopedia of consciousness* (pp. 279–293). Oxford: Academic.
Shepherd, G. M. (2011). *Neurogastronomy: How the brain creates flavor and why it matters*. New York: Columbia University Press.
Stevenson, R. J. (2009). An initial evaluation of the functions of human olfaction. *Chemical Senses, 35*(1), 3–20.
Teller, C., & Dennis, C. (2012). The effect of ambient scent on consumers' perception, emotions and behaviour: A critical review. *Journal of Marketing Management, 28*(1–2), 14–36.
Van Gulick, R. (2011). Consciousness. *The Stanford Encyclopedia of Philosophy* Summer 2011. from http://plato.stanford.edu/archives/sum2011/entries/consciousness

Varendi, H., & Porter, R. H. (2001). Breast odour as the only maternal stimulus elicits crawling towards the odour source. *Acta Paediatrica, 90*(4), 372–375.
Varendi, H., Porter, R. H., et al. (1994). Does the newborn baby find the nipple by smell. *Lancet, 344*(8928), 989–990.
Wright, L. (1976). *Teleological explanations*. Berkeley: University of California Press.

Conclusion: Comparing Olfaction and Vision

The primacy of vision over other modalities has led to a bias in our understanding of perception. The visuocentrism that dominates our thinking has very deep roots that are reflected in our philosophical vocabulary. We can *glimpse* or *see* the meaning of a proposition. Using a tactile instead of visual metaphor, we can also *grasp* it. After we grasped enough propositions, we will form a certain *view* or *outlook* on the topic. Ideas are talked about as if they are visual objects that are placed before us. They can be *viewed* from certain *angles*. Different *positions* can be taken with respect to a certain *view*. There are no comparable olfactory metaphors. Undoubtedly, this way of speaking has given vision a special status in philosophy of perception. Philosophy books have titles such as "Seeing and Knowing" (Dretske 1969) or "Philosophy and the Mirror of Nature" (Rorty 1981). Philosophers always have a "point of view", whether it is a logical point of view, a scientific point of view, or the view from nowhere.

I hope that this book will trigger a reexamination of the assumptions about perception that are built into the philosopher's vocabulary. I do not mean to deny that in humans vision is the most important and interesting modality. Humans happen to be active during the day, when the light from the sun allows us to perceive the space around us visually. If humans

A. Keller, *Philosophy of Olfactory Perception*,
DOI 10.1007/978-3-319-33645-9

were active at night, we might have evolved echolocation like bats, to navigate in the dark. Philosophy of perception is visuocentric because at some point in our evolutionary history our ancestors found it easier to find food and avoid being eaten during the day than at night.

Once our ancestors became active during the day, it was important to optimize visual perception, even if it meant to reduce further the usefulness of olfaction. The most important evolutionary event that shifted importance from olfactory perception to visual perception was the shift from walking on all fours to bipedalism. Our head and with it the sense organs for both vision and olfaction were moved up from the ground when we became bipedal. This further accelerated the evolution away from an olfactory creature to a visual creature. Close to the ground, down in the bushes, there is much less to see and much more to smell. However, at eyelevel (or noselevel) vision is much more useful than olfaction.

Understanding that the different ways in which humans perceive visual stimuli and odors have been shaped by the unique evolutionary history of our species leads to the realization that perception differs between modalities as well as between species. Human vision is different from human olfaction in interesting ways. Human vision is also interestingly different from fruit fly vision. Many of the features that are associated with one modality in humans are associated with another modality in other species. In humans, visual perception has spatial structure whereas olfactory perception does not. For us, colors can be, and usually are, arranged in a spatial pattern, while it is not possible for humans to perceive a spatial arrangement of smells. The presence of spatial structure in vision and its absence in olfaction are due to the structure of the human sensory systems. Nothing in the nature of olfactory perceptual qualities prevents them from being spatially arranged. It is easy to imagine an olfactory system that mediates spatially structured olfactory perception. Our whole body could be covered in odor-sensitive cells. We then would experience the smell stronger at the side of our body that is facing the odor source. This is how we perceive heat and why we can tell whether we walk toward or away from the sun based on the pattern of heat perception on our skin. Similarly, we cannot conclude that visual perception is inherently spatial from the fact that human visual perception has spatial structure. Many animal species can perceive visual stimuli, but do not have image-resolving eyes, and therefore do not perceive spatially arranged visual perceptual qualities. The simplest eye consists of only a single photoreceptor (Jékely et al. 2008). With such an eye, all that can be perceived is the level of illumination. A species of marine mollusks has hundreds of small simple eyes distributed over their body surface (Boyle 1969). They perceive a light source in the same way in which humans perceive a heat source.

In the light of this diversity of perceptual systems, generalizing from human vision to perception in general seems misguided. To make progress, other forms of perception have to be investigated. In this book, I have considered human olfaction. Throughout the rest of this conclusion, I will compare the central findings about olfactory perception with what we know about visual perception to show that many conclusions from the study of human vision do not generalize well to other modalities.

At the level of perceptual qualities, olfaction is more complex than other senses. There are more smells than colors or tones. Consequently, arranging the perceptual qualities of olfaction into a similarity space has not been possible yet. For colors or tones, this first step of systematically describing perception has been accomplished. However, once the perceptual qualities are individuated, olfaction is much less complicated than vision. What is perceived during olfactory perception in normal cases are individual olfactory perceptual qualities. The olfactory perceptual qualities are not arranged spatially and the temporal structure of odor perception is much impoverished compared to the other senses. This is in contrast with visual perception, which normally is the perception of visual scenes that are arrangements of color qualities in space. A philosophy of perception based on smells would focus heavily on the nature of perceptual qualities rather than on how they are arranged in time and space. Consequently, entities that consist of spatially and temporally arranged perceptual qualities, like perceptual events and perceptual objects, would not be part of perceptual philosophy if olfaction would be the paradigm sense.

The lack or relative unimportance of spatial and temporal information in odor perception also has consequences for our understanding of the relation between the physical world and our percepts. In vision, relative position and distance, as well as temporal sequences, in the physical world are often accurately represented through perception. This is adaptive because when the perceiving organism responds behaviorally to a stimulus, the responses have to be directed and timed based on the spatial and temporal structure of the physical world to have the intended effect on the physical world. The largely correct mapping of spatial and temporal similarities in the physical world onto spatial and temporal similarities in visual percepts has resulted in the idea of perception as a representation of the physical world. In olfactory perception, perceptual qualities are not arranged in space. When perception without spatial structure is considered, physical similarities between stimuli are not always mapped onto similarities in perception. In olfaction, physical similarity between odor molecules is not reliably mapped onto similarities between the perceptual qualities associated with the stimuli.[1] Because olfactory perception consists of the perception of perceptual qualities with no spatial structure, the relation between similarities in olfactory stimuli and similarities in olfactory perception is weak. If olfaction were the paradigm, the idea that perception represents the physical world would not have the same appeal it has when visual perception is considered. In olfaction, perceptual qualities are closer aligned with appropriate behavioral responses to the stimuli than with the physical features of the molecules. Olfactory perception illustrates that what perception does is not mirroring reality, but guiding behaviors.

Another striking difference between vision and olfaction is that olfaction is more difficult to demarcate than vision. While differentiating vision or audition from other forms of perception, olfaction is easily conflated with other chemical senses and involved in the multimodal perception of flavors. Individuating modalities based on representations, phenomenal character, the proximal stimulus, or the sense organs will lead to dramatically different results in the case of the chemical senses, while all four methods largely agree about how to individuate vision or audition. The difficulty of individuating olfaction may have contributed to its relative unpopularity among philosophers of perception. Some apparent disagreements in the philosophy of olfactory perception are a consequence of olfaction being individuated differently by different scholars.

In addition to the more fluid boundaries between olfaction and other sense modalities, olfaction is also difficult to distinguish from non-perceptual processes like emotion processing. Olfactory and emotional processing share the same neuronal substrate, and evaluative emotions such as disgust are intricately linked to odor processing. There is also widespread cognitive penetration of the olfactory sensory system. Olfaction illustrates therefore the interconnectedness of the mind better than many other modalities that have a different evolutionary history.

That the different evolutionary histories of the different modalities result in differences in how they are connected to cognitive processes is also reflected by the differential ineffability of the senses. Naming and talking about what is perceived through olfaction is more difficult than naming and talking about what is perceived visually. This also has far-reaching consequences. As speculated above, the perception that is most easy to talk about is the obvious candidate for philosophers to investigate. This may have contributed to the bias toward visual perception in philosophy. The strong connection between vision and language also has methodological consequences for the empirical study of perception. In visual psychophysics, the ability to name a stimulus is often used as a read-out. Often the reliance on verbal report is pushed so far that the absence of a verbal report is interpreted as the absence of perception, or at least the absence of conscious perception. In olfaction, the danger of drawing conclusions about perception based exclusively on verbal reports is less pronounced.

Taken together, an olfaction-based philosophy of perception would be much different from the familiar version based on visual perception. Where to go from here? One possibility is that the account of olfactory perception given here is seen as supplementing the well-established account of visual perception. The two accounts differ, but the differences between them do not amount to different suggestions about how we should think about perception. Instead, the differences between the account of visual perception and the account of olfactory perception reflect the fact that visual perception in humans is fundamentally different from human olfactory perception. An alternative to embracing the heterogeneity of perception is to try to identify features that are shared by all instances of perception. The differences between the perceptual systems are so pronounced that some are skeptic that a unified account that applies to all types of perception can be developed (Martin 1992). I am more optimistic and believe that enough features are shared between all forms of perception that an interesting general account of perception can be developed. The most striking similarities between olfaction and vision that the analysis presented in this book revealed are that both types of perception involve perceptual qualities and that they both serve to guide behaviors. This, I propose, could be a starting point for a modality-neutral theory of perception.

Note

1. To some degree, it is possible to predict the perceptual quality of a molecule based on its physical features. Molecules containing a sulfur atom, for example, have a distinct smell. It is also possible that future research will show that the relationship between physical features of molecules and perceptual olfactory qualities is closer than currently appreciated (for a review of structure-odor relationship research, see Rossiter, K.J. (1996). "Structure–Odor Relationships." *Chemical Reviews* **96**(8): 3201–3240).

References

Boyle, P. R. (1969). Rhabdomeric ocellus in a chiton. *Nature, 222*(5196), 895.

Dretske, F. I. (1969). *Seeing and knowing*. Chicago: University of Chicago Press.

Jékely, G., Colombelli, J., et al. (2008). Mechanism of phototaxis in marine zooplankton. *Nature, 456*(7220), 395–399.

Martin, M. (1992). Sight and touch. In T. Crane (Ed.), *The contents of experience*. Cambridge: Cambridge University Press.

Rorty, R. (1981). *Philosophy and the mirror of nature*. Princeton: Princeton University Press.

Rossiter, K. J. (1996). Structure-odor relationships. *Chemical Reviews, 96*(8), 3201–3240.

Illustration Credits

Figure 1	modified from original work by The Emirr (diagram of nasal cavity; in the public domain under the Creative Commons license (CC BY 3.0)) and Brad Ashburn (flower; The Noun Project)
Figure 2	modified from original work by SharkD (in the public domain under the Creative Commons Attribution-Share Alike license (CC BY-SA 3.0))
Figure 3	by the author
Figure 4	by the author
Figure 5	by the author
Figure 6	by the author
Figure 7	modified from original work by Nanobot (in the public domain)
Figure 8	by the author
Figure 9	by the author
Figure 10	(A) modified from original work by FrozenMan (in the public domain under the Creative Commons license (CC BY-SA 4.0)); B and C by the author

A. Keller, *Philosophy of Olfactory Perception*,
DOI 10.1007/978-3-319-33645-9

Figure 11	(A) modified from (Porter, Craven et al. 2007), with permission; (B) from (Jacobs, Arter et al. 2015) in the public domain under the Creative Commons license (CC BY)
Figure 12	(A) modified from the official MTA subway map; (B) modified from original work by SPUI (in the public domain under the Creative Commons License)
Figure 13	(A) modifies from original work ("Mimulus nectar guide UV VIS") by Plantsurfer (in the public domain under the Creative Commons license (CC BY-SA 3.0)); B by the author
Figure 14	by the author
Figure 15	by the author

Index

Note: Page numbers followed by "n" refer to notes.

A. Keller, *Philosophy of Olfactory Perception*,
DOI 10.1007/978-3-319-33645-9

CPSIA information can be obtained
at www.ICGtesting.com
Printed in the USA
LVHW07*1848080418
572701LV00017B/1737/P